The Wine
of Life
and other Essays
on Societies,
Energy &
Living Things

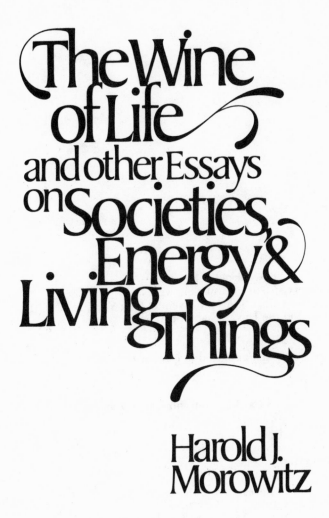

The Wine of Life
and other Essays on Societies, Energy & Living Things

Harold J. Morowitz

St. Martin's Press, New York

All of these essays first appeared in *Hospital Practice*.

Library of Congress Cataloging in Publication Data

Morowitz, Harold J
   The wine of life, and other essays on societies,
energy living things.

   1. Biology—Addresses, essays, lectures. 2. Social history—Addresses,
essays, lectures. I. Title.
QH311.M79   574   79-16404
ISBN 0-312-88227-0

TABLE OF CONTENTS

Preface

Without prior warning the mail one day produced a letter inviting me to write a column on various aspects of science for the magazine *Hospital Practice*. Owing to the indulgence of the publisher I have been presented with a mini-soapbox to stand upon and broadcast opinions to the world, or at least to the readers of one general medical journal.

Because the content of these pieces is, in effect, unrestricted, there is a challenge to continuously set down ideas on paper, a kind of intellectual put up or shut up. This exercise is recommended with enthusiasm. It cleans out mental cobwebs and induces a certain amount of healthy humility.

The project of writing articles, in a curious way, became a family endeavor as my wife and whichever children happened to be home at the time were transformed into editors, critics, and introducers of topics, ideas, and strongly worded opinions. Thus a number of cherished pieces ended up in the waste basket because they wouldn't pass my beloved critics. The dinner table became the boardroom as we debated great and small issues with equal fervor and occasional losses of our cool.

As time passed, the thought emerged that these essays might have a broader-based readership than the intended professionals. Articles were quoted in magazines and newspapers, and the idea of gathering these writings in a single volume developed. Musing over the collection I feel impelled to somehow characterize it, for my own benefit at least.

In looking at science, life, and my fellow human beings, my mind in an undisciplined way detects the cosmic within the nitty gritty and the trivial within the infinite. I believe that deep and important issues should be approached with sufficient good humor to keep us from regarding our mutable opinions as eternal truths. While not ignoring the real tragedy in the world, I feel it important to concentrate on hope. Given the existential dilemma of forever unanswered questions about our universe, I believe that joy is more fun than sadness and no further from the elusive reality of things. In short, it should be possible to be profound without being boring or afflicted with malaise.

Writing on just about everything has evoked more of the past than I ever envisioned was stored in memory. In particular I realize how attitudes and thoughts have been shaped by an undergraduate major in physics and philosophy. I also sense my debt to those mid-nineteenth century Americans —Thoreau, Melville, Poe, and J. Willard Gibbs —and have come to appreciate how deeply and differently I have been touched by Omar Khayyam, Socrates, Rabbi Hillel, Gautama Buddha, and Spinoza. The words of a number of undergraduate teachers have come vividly to life. I remember poetry read to me by my mother during a childhood illness. Recalled are the volumes of the Harvard classics which kept me company during the loneliness of early high-school years. Memories are revived of studying Bible with my grandfather. These are the kinds of things that curiously surface in writing of this nature.

If many threads weave through these essays, the fabric itself is a deep affection for science, rationalism, and that undefinable spark that makes us human and creative. The world is endlessly fascinating. I have picked items that excite me and attempted to share that excitement. I must confess that writing these has been fun. I hope that reading them yields an equal pleasure.

The Wine of Life

of Life

and other Essays

on Societies, Energy & Living Things

PART I

The Six Million Dollar Man—SCIENCE AND HUMANISM

The Six Million Dollar Man

Another annual cycle inevitably passed and the pain was eased by a humorous birthday card from my daughter and son-in-law. The front bore the caption "According to BIO-CHEMISTS the materials that make up the HUMAN BODY are only worth 97¢" (Hallmark 25B 121-8, 1975). Before I could get to the birthday greeting I began to think that if the materials are only worth ninety-seven cents, my colleagues and I are really being taken by the biochemical supply companies. Lest the granting agencies were to find out first, I decided to make a thorough study of the entire matter.

I started by sitting down with my catalogue from the (name deleted) Biochemical Co. and began to list the ingredients. Hemoglobin was $2.95 a gram, purified trypsin was $36 a gram, and crystalline insulin was $47.50 a gram. I began to look at slightly less common constituents such as acetate kinase at $8,860 a gram, alkaline phosphatase at $225 a gram, and NADP at $245 a gram. Hyaluronic acid was $175 a gram, while bilirubin was a bargain at $12 a gram. Human DNA was $768 a gram, while collagen was as little as $15 a gram. Human albumin was down at $3 a gram, whereas bradykinin was $12,000 for a gram. The real shocker came when I got to follicle-stimulating hormone at $4,800,000 a gram—clearly outside the reach of anything that Tiffany's could offer. I'm going to suggest it as a gift for people who have everything. For the really wealthy, there is prolactin at $17,500,000 a gram, street price.

Not content with a brief glance at the catalogue, I averaged all the constituents over the best estimate of the percent composition of the human body and arrived at $245.54 as the average value of a gram dry weight of human being. With that fact burning in my head I rushed over to the gymnasium and jumped on the scale. There it was, 168 pounds, or, after a quick go-round with my pocket calculator, 76,364 grams. Remembering that I was 68% water, I calculated my dry weight to be 24,436 grams. The next computation was done with a great sense of excitement. I had to multiply $245.54 per gram dry weight by 24,436 grams. The number literally jumped out at me—$6,000,015.44. I was a Six Million Dollar Man—no doubt about it—and really an enormous upgrade to my ego after the ninety-seven cent evaluation!

Assuming that the profits of the biochemical companies are considerably less than the 618,558,239% indicated above, we must still strike a balance between the ninety-seven cent figure and the six million dollar figure. The answer is at the same time very simple and very profound: information is much more expensive than matter. In the six million dollar figure I was paying for my atoms in the highest informational state in which they are commercially available, while in the ninety-seven cent figure I was paying for the informationally poorest form of coal, air, water, lime, bulk iron, etc.

This argument can be developed in terms of proteins as an example. The macromolecules of amino acid subunits cost somewhere between $3 and $20,000 a gram in purified form, yet the simpler, information-poorer amino acids sell for about twenty-five cents a gram. The proteins are linear arrays of the amino acids that must be assembled and folded. Thus we see the reason for the expense. The components such as coal, air, water, limestone and iron nails are, of course, simple and correspondingly cheap. The small molecular weight monomers are much more complex and correspondingly more expensive, and so on for larger molecules.

This means that my six million dollar estimate is much too low. The biochemical companies can sell me their wares for a mere six million because they isolate them from natural products. Doubtless, if they had to synthesize them from ninety-seven cents worth of material they would have to charge me six hundred million or perhaps six billion dollars. We have, to date, synthesized only insulin and ribonuclease. Larger proteins would be even more difficult.

A moment's reflection shows that even if I bought all the macromolecular components, I would not have purchased a human being. A freezer full of unstable molecules at −70° C does not qualify to vote or for certain other inalienable rights. At six billion it would certainly qualify for concern over my −70° deep freezer, which is always breaking down.

The next step is to assemble the molecules into organelles. Here the success of modern science is limited as we are in a totally new area of research. A functionally active subunit of ribosomes has been assembled from the protein and RNA constituents. Doubtlessly other cellular structures will similarly yield to intensive efforts. The ribosome is perhaps the simplest organelle, so that considerably more experimental sophistication will be required to get at the larger cell components. One imagines that if I wanted to price the human body in terms of synthesized cellular substructures, I would have to think in terms of six hundred billion or perhaps six trillion dollars. Lest my university begin to salivate about all the overhead they would get on these purchases, let me point out that this is only a thought exercise, and I have no plans to submit a grant request in this area.

Continuing the argument to its penultimate conclusion, we must face the fact that my dry-ice chest full of organelles (I have given up the freezer, at six trillion it simply can't be trusted) cannot make love, complain, and do all those other things that constitute our humanity. Dr. Frankenstein was a fraud. The task is far more difficult than he ever realized. Next, the organelles must be assembled into cells. Here we are out on a limb estimating the cost, but I cannot imagine that it can be done for less than six thousand trillion dollars. Do you hear me, Mr. Treasury Secretary, Mr. Federal Reserve Chairman? Are these thoughts taking a radical turn?

A final step is necessary in our biochemical view of man. An incubator of 76,364 grams of cell culture at 37°C still does not measure up, even in the crassest material terms, to what we consider a human being. How would we assemble the cells into tissues, tissues into organs, and organs into a person? The very task staggers the imagination. Our ability to ask the question in dollars and cents has immediately disappeared. We suddenly and sharply face the realization that each human being is priceless. We are led cent by dollar from a lowly pile of common materials to a grand philosophical conclusion—the infinite preciousness of each person. The scientific reasons are clear. We are, at the molecular level, the most information-dense structures around, surpassing by many orders of magnitude the best that computer engineers can design or even contemplate by miniaturization. The result must, however, go beyond science and color our view of the world. It might even lead us to Alfred North Whitehead's conclusion that "the human body is an instrument for the production of art in the life of the human soul."

All the World's a Stage

From the dawn of civilization until a few hundred years ago, most Western thought regarded existence as a cosmic drama being played on a small stage for a relatively short time. Each society had its creation myth, Act I, Scene I; each script had its heroes and villains. Each culture had its producers or directors who ordered the lights to be turned on, instructed the actors, and reacted with anger when the performances did not go well. Whether we read biblical narrative, classical plays of Greece, *The Thousand and One Nights*, Boccaccio's *Decameron*, or Shakespearean plays, the same sense of drama persists concomitantly with the importance of the individual. The universe was a newly built stage, the script was fresh, and every bit player was important.

Other voices were heard in the world. Vedantic mystics embedded themselves in a sense of timelessness and Buddha set forth a radical rejection of metaphysics in return for a religious existentialism which sought to alleviate the pain of existence. And even in the West occasional voices doubted that the play had a beginning or an end or could be understood by the cast. The enigmatic book of Ecclesiastes noted, "All the rivers run into the sea; yet the sea is not full; . . . That which has been is that which shall be; . . . And there is no new thing under the sun." In a like vein, Omar Khayyam would write more than a thousand years later, "We are no other than a moving row/of visionary shapes that come and go."

Nevertheless, the idea of life as divine drama persisted and animated the Renaissance and succeeding generations. The modern malaise or loss of sense of the drama is not that of Ecclesiastes or the East but of more recent origin, beginning with spirited men like Copernicus (1473–1543), Descartes (1596–1650), and Spinoza (1632–1677), whose own lives were antithetical to a mood of malaise. The first of these, by his studies in astronomy, moved the stage from the center of the universe off into one of the side rooms. The second got us to doubting what we were seeing, and the third introducer of modern thought took the creative force out of God's role, reducing it to a set of inexorable, timeless laws. The great mass of Western culture was relatively uninfluenced by these natural philosophers and the older view of the unfolding scenario dominated up until quite recent times.

The century from 1830 to 1930 was the most dynamic period in altering and reshaping mankind's world view. In astronomy the development of stellar catalogues reduced our sun to the status of one medium-sized star in a galaxy containing a vast number of such stars. Finally, the galaxy also was found to be one of many; by 1908, 13,000 star clusters and nebulae had been catalogued. Geological studies made it clear that the earth was much older than the biblical five or six thousand years. In 1864, William Thompson reasoned on the thermodynamic grounds that the earth must be at least 20 million years old. In 1905, radioactive dating was introduced and rocks of ages of over two billion years were found, finally leading to the concept of a 4.6-billion-year-old planet. Our act was now being played out in a very tiny, extremely ancient theater almost lost in an expanding universe.

The mid-1800s brought at least two other major disappointments to man's self-image. First, the second law of thermodynamics announced that the universe was running downhill and in time would expire. Then Darwinian evolution suddenly severed our ties to the angels and presented us with an entire new array of hairy relatives whom we did not like all that well. While science and technology were providing brilliant insights into our world and easing our everyday life they were simultaneously depriving us of a world view in which our importance was assured. The dynamic century gave us many things, yet took away much that was also greatly valued.

All of the preceding developments are now moving from the scholars' desks into the field and the malaise spreads to all of literate society. For many the drama is over. We are envisioned as impotent monkeys existing for but a flicker of time in a tiny insignificant corner of the galaxy; how different in tone from Abraham arguing with God over the morality of Sodom and Gomorrah or the voice out of the whirlwind asking of Job, "Where wast thou when I laid the foundations of the earth?"

There is no doubt that malaise has become a social disease of our time. College students, my young friends, assure me that the pathology is real and widespread and painful. This condition can perhaps best be seen in the novels of Saul Bellow, who has become a Nobel Laureate in literature by making the self-pity of the anti-hero a central theme of the modern novel. The reviewers tell us that the show is over and the actors are destined to sadly spend their time, tin cup in hand, waiting in line at the cosmic soup kitchen.

But wait a moment before ringing down the curtain! Surely somewhere in the core of our beings we must sense that the pathology is not fatal and it too will pass. This malaise comes from a one-sided reading of the laws of nature. We know that each time a sperm and egg fuse, a combination of genes occurs which is unique in the history of the universe. The potential for creativity and innovation is inexhaustible. Our species may itself evolve to some undreamed-of potential. We have the capacity to reach out to the stars and if we have only touched the moon, our efforts in this direction are very new. The exploration of inner space, the mind of man, is also only at the barest beginnings. Even if the earth is extremely old, civilization is refreshingly new, hardly more ancient than the fundamentalists' chronology of the planet.

Malaise and self-pity lead to self-fulfilling prophecies. If we wallow in the feelings that all is despair, we will not do that which must be done to dispel the cloud of despair that has descended. We must lose patience with philosophical pessimism. If our global self-image was wrong it must be discarded and efforts begun to craft a more believable script. In that rewrite job we must remember to include not only atoms and molecules and photons but also our creative thrust, and most of all our capacity to love.

If the universe is infinite, then even Copernicus would have had to agree that the stage on which you stand is the center of that universe. If the universe is very old, then even Spinoza would have had to admit it took a long time to invent a character as unique as yourself. If the universe is disturbingly hard to comprehend, then even Descartes would surely have conceded that a challenge worthy of your mind is set before you. If death is mysterious and terrifying, then you must ask if it is really too high a price to barter for the potentialities of existence.

Therefore we need simply to ignore those critics who would close down the show; they have missed the message—it has gone over their heads. All the world is still a stage and it is time to work on the next scene, attempting to perfect every aspect of our performance. The show must go on.

On Riding a Biocycle

Often we seize upon a concept that seems new and fresh at first glance but upon reflection recedes deep into antiquity. The notion of nutrient cycling in global ecology is just such an idea. Under the misnomer of "recycling," the idea of nutrient flows has become the password of a generation of environmentally concerned individuals who frequently act as if the notion of biodegradability was discovered sometime late during the Johnson administration.

However, one hundred years ago the poet Walt Whitman had elaborated the theme in exquisite, if gory, detail in his poem, "This Compost," which ends:

Now I am terrified at the Earth! it is that calm and patient,
It grows such sweet things out of such corruptions,
It turns harmless and stainless on its axis, with such endless
    succession of diseased corpses,
It distills such exquisite winds out of such infused fetor,
It renews with such unwitting looks, its prodigal, annual,
    sumptuous crops,
It gives such divine materials to men, and accepts such leavings
    from them at last.

The thought is a recurring one in Whitman's poetry and

11

emerges in his work "Grass" in the optimistic philosophy, "The smallest sprout shows there is really no death."

Ideas of nature may be scientific or philosophical but they are also the stuff that poetry is made of. Since both the poet and the scientist are trying to comprehend the same natural world, it is hardly surprising that their thoughts should overlap. Thus John Milton's poetry labors with the problems of cosmology and wavers between Copernican, Ptolemaic, and biblical views of the nature of the solar system and problems of celestial mechanics. In the first century B.C., Titus Lucretius Carus combined scientist and poet in the same individual so that his famous essay, "On the Nature of Things," is in fact one long poem. It is also the major scientific treatise of that period.

Returning to the idea of material cycling on the surface of the earth, we can see that it has an even longer history than just indicated. Three hundred or so years before Whitman, William Shakespeare, in the play *Timon of Athens*, had perceived the same theme and penned the fascinating lines:

> . . . : *the earth's a thief*
> *That feeds and breeds by a composture stolen*
> *From general excrement.*

It would be hard today to outdo the bard in this terse description of the process.

Five hundred years before Shakespeare, the Persian sage Omar Khayyam had written the lines that Edward Fitzgerald has translated:

> *I sometimes think that never blows so red*
> *The Rose as where some buried Caesar bled;*
> *That every Hyacinth the Garden wears*
> *Dropt in her Lap from some once lovely Head.*

Indeed the whole of the *Rubaiyat* is a plea for life before each individual is himself cycled back into the nutrient pool.

These biological facts are not only there for Omar; they became the source of an imperative as to how life should be lived. Thus:

> *Ah, make the most of what we yet may spend*
> *Before we too into the Dust descend*
> > *Dust unto Dust, and under Dust to lie*
> *Sans wine, sans song, sans Singer, and—sans End!*

Omar was a scientist, best known to his contemporaries as an astronomer and mathematician. His quatrains somehow combine science, philosophy, and poetry into a coherent if somewhat pessimistic whole.

We can move back into history one thousand years before Omar when the Talmudic scholar Hillel used to say, "The more flesh the more worms." In the same book of the Talmud, *The Wisdom of the Fathers*, we find Akabya Ben Mahalalel quoted as saying, "Whence are thou come? From a putrid drop. Whither are thou going? To a place of dust, worm, and maggot." Hardly a romantic description, but it is not easy to be romantic about detritus.

Back into the past before the Talmud to the earliest record of the Book of Genesis we find Adam cursed:

> *In the sweat of thy face shalt thou eat bread*
> *Till thou return unto the ground*
> *For out of it wast thou taken*
> *For dust thou art*
> *And unto dust shalt thou return.*

Of course the notions of cycling could not fully have moved from the realm of poetry to the realm of science in the premodern period. Indeed the formulation of the principle of conservation of mass following Lavoisier's experiments in the late 1700s was a prerequisite to the realization of the formal nature of cyclic flows. The modern theory of the carbon and nitrogen cycles was then introduced by Justus von Liebig in the mid-1800s. The first formal statement came in a book published in 1840 entitled *Organic Chemistry and its Applications to Agriculture and Physiology*. Subsequent editions of that work as well as the work of others considerably refined the scientific theory of nutrient cycling.

These views soon achieved their place in biology textbooks, but did not become issues of public interest until the recent past, when the full impact of the finite nature of the planet impressed itself on the mind of the body politic.

As the science of bioenergetics emerged, it became clear that the continuous flow of matter from soil, air, and water into biomass and then back to soil, air, and water is a process driven by the energy of the sun. Cycles occur because the energy of photosynthesis leads to a buildup of complex structures, whereas the catabolic processes ultimately related to the second law of thermodynamics are associated with the breakdown of biomass and the return to the nutrient pool. The existence of cycles is thus deeply rooted in the underlying physical process of energy-flow and is tied to the cosmic process by which our star formed and now dissipates its radiant energy into space.

A famous biologist has been rumored to have said, "I'm only interested in wine, women, and song, and all three are biodegradable." This *bon mot* is a curious echo of Omar's philosophy some ten centuries earlier. The ideas of nutrient cycling are perhaps moving back into poetry. Maybe that's the route of a really noble idea—from poetry to science and back to poetry. C. P. Snow's two cultures need not exist in isolation.

Sole on Fire

How are we to deal with observations that appear to lie outside of the realm of normal scientific explanation? If they are recorded as happening only a few times or not occurring in the presence of trained observers, we are free to reject them as not being fit subject matter for science. However, if they are frequently reproducible under a wide variety of conditions, they then stand as a challenge to some particular branch of science until they can be fitted into an explanation or until the theoretical structure is altered to include the anomalous observations.

Fire walking appears to present the kind of challenge to biophysics that is indicated above. The practice is known in a number of cultures in East Asia, Pacific Oceania, and certain other societies. Two forms of the act seem to be followed: walking on fire-heated stones and walking on a bed of burning charcoal. Each of these is practiced by separate groups in Fiji, and subsequent remarks will be directed at Fijian fire walking, since I have personally observed one kind and have direct reports from reliable observers on the other.

Walking on hot stones is practiced in Fijian culture by a select group of men from a few villages on the island of Beqa (pronounced Mbenga). The fire walkers claim descent from a common ancestor who, according to legend, was given power over fire as a reward for releasing a captured demi-god who was in the form of an eel when captured. The fire is built in a pit about six meters in diameter and consists of large logs and kindling wood. Mixed in among the combustibles are rocks about 20 to 50 centimeters in size. The fire is lit several hours prior to the ceremony, which has a ritual character. After the burning period, the uncombusted portions of the logs are removed from the pit and the hot rocks are spread out in a uniform layer. The participants then walk in single file, barefooted, across the pit. They may pause briefly in the pit or squat momentarily. There appears to be little in the way of accurate measurement of the temperature of the rocks, but stones prepared in this way are used in underground ovens for roasting foods.

Sir Thomas Henley gives the following report (T. Henley, *Fiji, the Land of Promise*, John Sands Co., Sydney, 1926). "Medical men have stated that they have examined the feet of the performers before and after walking on the stones and although the thermometer registered between 300 and 400 degrees (Centigrade) the fire had not affected them. I was advised that a European, some years back, tried it with his shoes on and burnt his feet so badly that he had to be carried off the Island."

Fire walking among the Indian community of Fiji is carried out as a Hindu religious ceremony. Any of the faithful can participate following a two-week period of purification, prayer, abstinence, and special diet. The rite is performed under the direction of a priest and forms part of a more extensive religious event. A shallow pit is built some four or five meters on a side and 15 to 20 centimeters in depth. A fire of mangrove wood burns in this hole in the ground for many hours prior to the ceremony. The burning coals are then raked out in a flat bed and the devotees, in an ecstatic state and under instructions from the priest, walk across the bed of fire. The radiant heat from the bed of coals is sufficiently intense that an observer has reported to me that the nonparticipant faithful who are watching the rite cannot remain much closer than seven to ten meters. Again there seems to have been little done in the temperature measure although a related case of walking on a bed of coals has been reported (O. Degener, *Naturalist's South Pacific Expedition, Fiji*, Paradise of the Pacific Ltd., Honolulu, 1949).

"A Kashmir Indian, Kuda Bux, performed a similar stunt in London under the auspices of several scientists. In this case Mr. Bux, whose feet were not thick skinned, walked back and forth across a trench filled with burning charcoal having a surface heat of 800 degrees (Fahrenheit). When doctors examined Bux's feet after the ordeal, they not only found them uninjured but one fifth of a degree cooler than before the walk. An Englishman then took off his shoes and socks and tried to imitate Bux's performance. After two floundering steps on the charcoal, the Englishman left from the trench with his feet so badly burned and blistered that he was placed under a nurse's care."

The most detailed reports of temperature measurement during fire walking relate to a demonstration carried out in Honolulu in 1949 (Charles W. Kenn, *Fire Walking from the Inside*, Franklin Thomas Publisher, 1949). The reported surface temperature of the stones averaged 610°C as measured with thermocouples. The contact time per step was measured as three fourths of a second.

Given the somewhat meager information available on fire walking we may form four possible hypotheses:

a) When the phenomenon is exhaustively studied in terms of heat capacities, thermal transfer coefficients, properties of the skin, and other known factors, then no anomaly will exist and fire walking will be understood within the framework of current physiology, thermodynamics, and biophysics.

b) The participants do not actually fire walk, but an element of fraud is involved in the exhibitions; a sleight-of-hand or, more properly, a sleight-of-foot is used.

c) There are aspects of psychophysics that lie outside of and indeed are inconsistent with the present paradigm of biophysics. This would involve a reformulation of much of science.

d) The true explanations are the legends and theological statements that lie totally outside the syntax of science.

From the available evidence it seems possible virtually to eliminate the second hypothesis, that of fraud. Fire walking is, after all, a social phenomenon rather than a trick practiced by a few individuals. It is widespread historically and has presumably occurred independently among peoples at different times in history. The approach of the religious devotees is completely incompatible with a fraudulent attitude, as the reasons for fire walking are deeply intertwined with profound personal desires and fears of the walkers.

That leaves us with three possibilities: one firmly within conventional science (a), one opening up new and fascinating aspects of science (c), and the third lying totally outside of science (d). The first possibility is certainly subject to detailed experimental study. Indeed a body of knowledge currently exists on conditions of burn injury, and this material in conjunction with physical studies on the fire pits should reveal the magnitude of the anomaly if it exists. Yet apparently very little of such research has been done, and the limited studies have started out with the assumptions that no challenges exist (see, for example, P. S. Langley on fire walking in Tahiti in the 1902 Smithsonian Institution Report). However, the following question may be asked: How many of those who maintain that no extraordinary problems are raised by the phenomenon of fire walking would be prepared to follow the Beqa villagers or the Hindu devotees out across the hot pits? The answer to this question clearly indicates that we cannot subsume fire walking under the paradigm of conventional science without first subjecting it to extensive experimental investigation.

As Western science interacts with conceptualizations from other cultures, a number of overlaps occur. Folk medicine from around the world has provided pharmaceuticals derived from plants and therefore fits into a conventional framework. Other areas such as acupuncture are trouble-some because we lack a complete neurophysiological model to account for the observations. Fire walking appears at first sight to be even farther from the accepted framework. It is, however, reproducible and therefore open to experimental investigation.

Until the answers are all in we must remember Hamlet's advice to his friend. "There are more things in heaven and earth, Horatio, than are dreamt of in your philosophy."

Bay to Breakers

Perhaps it is our obsession with getting away from the "rat race" or maybe it is some more primordial urge hiding deep within us, but there can be no doubt about the fact: Americans have taken to running. This bit of the obvious was riveted firmly into consciousness as I looked around at 12,000 fellow runners crowding onto the pavement near the Embarcadero in San Francisco. The occasion was "Bay to Breakers," a celebrated 7.6 mile race from the foot of the Bay Bridge across the city to the finish on the shores of the Pacific. Through the filter of the imagination everyone in the crowd seemed sleek and athletic, and the cerebrum was tickled by a sense of "What am I doing here?" My life began to flash before me, or at least my life as a runner, those 12 months of untiring effort, pain, and exhaustion that preceded the ultimate challenge that now lay ahead.

It all began innocently enough, jogging around the track in an effort to magically dissolve away a few unwanted fatty acid molecules and to convert a figure, transmogrified by too many years of sitting at a desk, back into the sylphlike Olympian form I thought I remembered from my youth. A few weeks of this activity induced an inordinate pride, a few careless remarks, and too soon the gauntlet was thrown down: "Will you run the Bay to Breakers next year?" The results were predictable; machismo overwhelmed common sense and the rest became history. Winter days in New Haven involved slogging through snow to get to the gymnasium to run circles around a small indoor track. A vacation in Honolulu included puffing up the side of Punchbowl. And now at last the moment of destiny.

But thoughts of the past are not suitable to the commencement of a great enterprise, and the mind turns inexorably toward the deeper, more profound aspects of the existential present. The act of running is linked to that subtle balance of energy by which we ingest aliment, maintain our thermodynamically unstable systems, and metaphorically speaking, engage in the chase, our ultimate animal function. The final step in each and every one of these biological processes is the transformation of energy from adenosine triphosphate (ATP) to its ultimately usable form. In the muscle this conversion changes the chemical energy of ATP into mechanical work directed along the actomyosin fibers. It is a process not yet completely understood, yet one that makes possible every motion we undertake. The brain wills and the muscle contracts; it is the physiological essence of the mind-body problem. Doubtlessly one of these days a poet will appear to honor the molecule of ATP. Until then we can only celebrate this vital part of us in the somewhat cold, formal language of biochemistry. As a well-known sports announcer would have put it: "ATP is the most underrated molecule in the league today."

All of these ramblings were preparing the psyche for the ordeal ahead. I turned to my running companions and announced: "Think ATP." At that moment there occurred one of those wondrous coincidences that elevate life from the ordinary and constantly lead us to thoughts of the parapsychologic. A face in the crowd appeared to be that of one of the world's experts on the interaction of ATP with actomyosin. Knowing him only slightly and being struck by the strangeness of the situation, I refrained from checking his identity. Weeks later, meeting him at a seminar, I did indeed confirm his presence at the start of the race that morning. The explanation was less than mystical; his wife was running and he had driven her to the staging area.

In any case, again musing about how to mobilize ATP, I was awakened from my reveries by a wave of excitement that sped through the crowd as 12,000 souls lurched forward and began to stampede down Howard Street. To describe the ensuing minutes as a race is to miss the essence of the event. The Boston Marathon is a race, but for San Francisco the Bay to Breakers is a happening. It is true that among the first thousand or so runners there is competition to cross the finish line before the other contestants. Back in the pack there is a different ambience; ordinary individuals are competing with themselves to meet a challenge that is near to the ultimate in their physical capabilities. It is a good-natured crowd engaging, at the start, in some amount of friendly repartee.

As we jog past First Street, thoughts of striated muscles are triggered by slight pangs in the legs. The mystery of muscles, as has been suggested, is how the potential stored in ATP molecules gets released and causes a sliding between parallel actin and myosin molecules leading to the contraction of the fibers. The mechanism is a direct chemomechanical conversion and is quite unlike familiar engines that involve a thermal intermediate. Since the beginnings of the industrial revolution, devices for the production of work from stored chemical energy have relied on the heat of the reactions to expand gases and drive pistons. Nature has chosen another course, the isothermal molecular engine, which converts chemical bond energy directly to mechanical work. When this mechanism is understood we will probably be able to design and build highly efficient molecular engines on an industrial scale. As we move along the course, thoughts on ATP power are orchestrated by the constant footbeats of the masses of runners. And then, before one can say "chemical potential," we have reached the first turn onto Ninth and across Market Street where a fire engine is honking its greeting and encouragement to the passing runners. Just past this intersection the course turns onto Hayes Street where a fantastic sight greets the avid participants. Up the hill as far as the eye can see is a solid cohort of runners expending their maximum output as they struggle against gravity. The sidewalk and the road are filled with sweating humanity, and the sound of footbeats is drowned out by the cacophony of heavy breathing. Halfway up the hill we pass a "runner" in a wheelchair, and a wave of applause breaks out as his fellow participants spontaneously empathize. On Hayes Street this morning one can feel optimistic about the human condition.

From the top of the hill the course winds down along the panhandle into Golden Gate Park. With about three miles to go, the months of training begin to tell as a number of challengers drop out of the run. Into the park and all thoughts of ATP fade into the background with the refreshing change from the city streets to the natural beauty of the lawns and gardens. Biochemistry gives way to field biology as we take strength from the trees and shrubs, and one senses a "second wind" invigorating most of the runners. After passing the pond, a cool sea breeze suddenly signals that the Pacific cannot be far away. Around the curve the windmill is in sight. We realize we are going to finish, and a "high" feeling animates the dash to the final line.

It is hard to know why it feels so good to have run the race. It is no great athletic feat, and the triumph is shared with 10,000 others. Yet there is somehow that special quality of having tested oneself. By now more glycogen is breaking down, the mitochondria are manufacturing ATP in large quantities, the muscles are reacquiring their energy store, and I walk along the shoreline searching for meaning. Subversive thoughts come to mind, including glimmers of next year's race. Perhaps I can cut five minutes off my time. Maybe I should get a T-shirt that says "ATP Power." There's just no limit to the spirit of man.

De Motu Animalium

We habitual almanac readers have a heavy burden to shoulder. Somehow our friends seem to get unduly impatient when we point out that Dusseldorf Airport had 3,031,284 international passengers in 1972, or alternatively that the Swiss merchant fleet weighed in at 219,000 gross tons in 1973. But turn us loose on sports records and it's a different story; almost everyone cocks an ear to hear about the possibility of running a mile in less than 3 minutes and 40 seconds. Appropriately enough, my book of facts nobly entitles one table "Evolution of the World Record for the One Mile Run."

And so it is now that the last Summer Olympics are part of history, we may go back and examine the record books to see what lessons can be gleaned from the times and trials of the world's great athletes since 1896. In every event the distances and times have shown steady improvement since that group of athletes met in Athens 80 years ago. The 100 meters, run in 12.0 seconds at the first Olympiad, was done in 9.9 seconds in 1968. The time for 1,500 meters has dramatically fallen from 4 minutes 33 seconds to 3 minutes 36 seconds while the 16-pound hammer throw has gone from only 167 feet 4 inches to over 247 feet in the 1972 competition. In every event, today's competitors produce routine scoring that would have seemed visionary at the turn of the century. In track, field, and swimming events alike, there has been a regular and steady advancement of performance over the years.

However, just musing over the records is not good enough. These are numerical results and we need to turn more powerful tools of mathematics on them to extract the full meaning and significance. The obvious initial approach is to graph the winning data versus the year for each event. This first bit of analysis produces some fascinating regularities. All the charts are scatter plots around straight lines of improving results. This simple observation is startling because the linear regression (straight line character) is really quite unanticipated. To detail the degree of our surprise, consider the most classic of Olympic events, the discus throw. In 1900 Rudolf Bauer won with a toss just over 118 feet. One would have expected that with better training and technique the distance would have risen rapidly at first and then leveled off at some point near the human limit for this event. But contrary to this expectation, the value has risen more or less linearly at something over one foot per year to the present record of 221½ feet set by Mac Wilkins in 1976. The fact that all of the records appear to have a linear character suggests that we have not come close to possible physical bounds in any of these events. This path of reasoning inevitably leads us to the question of what determines the maximum of man's prowess.

The theoretical studies of mammalian motion go back to the first biophysicist, the remarkable Italian scholar Giovanni Alfonso Borelli (there's poetry in that name) whose classic *De Motu Animalium* (On the Motion of Animals) was published in 1681. Borelli studied the mechanics of motion of a wide variety of fauna including man. The present state of this field may be found in a work like *Animal Locomotion* by James Gray, who notes: "There seems little doubt that the highest speed which can be reached as the result of a relatively brief but intense effort is that at which the total effective output of mechanical power developed by the muscles is utilized for keeping the limbs moving relative to the body, and periodically raising the center of gravity of the animal. The full output of mechanical power from the muscles can only be maintained if an adequate supply of oxygen is available, and the power of endurance, therefore, depends much more on the efficiency of the heart and lungs than on the intrinsic properties of the locomotory muscles." Thus, we seem to lack a well-defined discipline for attempting to predict with precision the human limitations that arise from biological factors.

On second thought, we should have perhaps realized that physiological considerations alone could not have determined the best that we are capable of. What has become quite apparent over the years is the great importance of psychological factors in all types of athletic events. The will to win cannot be ignored. The constraints at the moment are not of the human body but rather of the human spirit.

There is good news, therefore, in all the linear regressions and best fits of data obtained from these quadrennial competitions, for they stand in opposition to the ideas of a number of contemporary thinkers who have suggested that human progress is coming to an end and we are entering an age where totally different values will emerge. The most scientifically rooted of these, Gunther Stent, points "to the end of progress" (see *The Coming of the Golden Age*, Natural History Press, New York, 1971). He argues that this past age has been characterized by Faustian man reaching for the infinite and never being satisfied. Economic security, he goes on to say, weakens the drive of Faustian man and leads to more inner directed cultures. He further maintains that art and science have exhausted their domain and will also end. His future vision is a utopia of being and doing, free from creating and struggling. Whether one regards Stent's vision as paradise or inferno, one finds in the sports records a refutation of the basic postulates of the end of progress that underlies his essay.

A basic optimism emerges from the following sort of argument. The linear time dependence of performance implies that we are so far from human limits that we cannot even calculate from the curvature of the data where the final records will be. If we have not come close to the ultimate in our physical ability, how much further from our best must we be in our artistic and scientific development, which are much more recent activities of *Homo sapiens*. If we are far from the boundaries of our technical achievements, how much further must we be in the more humanistic and spiritual aspects of thought where the issues have barely been defined.

Rather than the cheerless boundaries envisioned by Stent, we rise to Olympian heights with Berton Brayley's cheerful verse:

> *With doubt and dismay you are smitten,*
> *You think there's no chance for you, son?*
> *Why the best books haven't been written,*
> *The best race hasn't been run.*

On Computers, Free Will, and Creativity

The advent of computer technology has raised a wide variety of issues relating "artificial intelligence" to human capabilities. Indeed, this branch of computer science has several practitioners and is a well-recognized field of study. Quite a number of years ago, in the early days of electronic calculators, the issue was raised as to whether or not computers could have free will. Although such questions obviously take us out of science in the narrower sense, their exploration may illuminate the meaning of difficult concepts. The issue of independent computer activity was treated fictionally in the movie *2001*, and the theme has appeared frequently in science fiction. Our dealing with this matter will be somewhat less dramatic but nevertheless has some surprising features.

To make the discussion more concrete, consider a computer that is programmed to play chess in the following manner. Before each move a subprogram lists all the possible moves that can be made within the rules of the game. A second subroutine evaluates each of the possible moves and assigns a value to each one based on its tactical advantage and the long-range strategy. The machine picks the choice of highest numerical advantage and makes its move. At the end of each game a third subroutine examines the result and reprograms the evaluation subroutine to incorporate what has been found out from the experience of the game. Thus far, the system has the following properties: a) it is rational—that is, it follows the rules of the game (an irrational computer would try to have a knight make a bishop's move or some other step that is not allowed) and tries to win; b) it is capable of learning (changing the values of the evaluation subroutine after each game, in response to the results of that game, is clearly a form of learning).

One more feature can now be added to the program. At the stage where the machine selects the move of highest tactical advantage, the program will ask whether there are additional moves with numerical advantages almost as great as the highest-valued one. If other moves are within some small numerical range of the top value, the computer will not automatically select the greatest one but will choose among the top ones by some random procedure. This process can consist of throwing dice, consulting a random-number table, or any other method that assures randomness. We will designate as epsilon the range of numerical values within which the random choice is made.

This act of introducing a small degree of randomization in the program now produces the property that the machine is incompletely determined. An observer on the outside, watching the machine play chess, will become aware of an entity that is rational, capable of learning, and yet incompletely determined, by any procedure. According to all the usual notions, an outsider would have to conclude that the machine has free will, since it pursues its goal of winning at chess, yet its moves cannot be predicted with certainty.

Altough the previous exercise may seem a strange approach to the problem of free will, it basically does affirm that any behavioral trait that can be defined operationally can be analogued by a machine. The philosophical sophistication must come in the operational definitions that we have assumed in a rather straightforward way. The example of the chess-playing machine has forced us to define free will in terms of observables, and this is always a useful procedure. Indeed, any abstract problem can be brought to a more tangible state by the use of operational definitions and modeling (at least in principle) of those operations with hardware. This general approach is a contribution of computer science to other branches of human thought that we are beginning to appreciate fully. In order to interface with a computer we must reduce a problem to precise symbolic elements. This reduction is often a major step in seeing the issues with greater clarity.

The example of the chess program has another aspect that is worth discussing. The quantity we called epsilon (using usual mathematical notation) designates the range of values over which the program makes random decisions. A program with a very small epsilon approaching zero will almost always choose the highest value decision and will be a good, conservative, almost deterministic chess player. It will approach the behavior of the system without the randomization step. A program with a large epsilon will make wildly unpredictable moves and will occasionally execute a brilliant maneuver. However, such a program will generally make poor moves and will usually play a bad game of chess. At some intermediate value of epsilon the machine will play a good game of chess and will imaginatively explore new strategies so that in time it will play a very different game than was in its original program. It is of course difficult to know the optimal value of epsilon.

This discussion of chess-playing machines has a certain analogy to creativity in humans. An individual with a very small epsilon always follows the existing program and plays "by the book." Such an individual may be very successful in the right pursuits, but cannot be described as creative in any sense. A person who operates with a very large epsilon may seem very imaginative, but rarely succeeds because his ideas are either too far from reality to work or too far from the accepted norms to make contact with others. A really large epsilon denotes a "crackpot."

Somewhere between these two extremes lies the golden mean, that value of epsilon that makes an individual open to a search for new ideas while keeping firmly rooted in reality, the art of the possible. It is in this domain that we find our creative geniuses. They possess the ability to explore creatively while at the same time they remain rooted in the domain of valid human experience.

An analogous situation appears to exist in the performing arts, where one can distinguish between excellent technicians, sloppy players, and brilliant performers. The technician plays the music exactly as it is written, with no deviation on the part of the performer from the program (the written music). The sloppy player wanders so far from the preassigned values that one feels that the performance is not true to the composer. In between these extremes is an epsilon that designates the performing artist who remains within the constraints imposed by the written notes, yet creatively explores the possibilities within these constraints. The great performer must be sufficiently disciplined by technique, yet sufficiently free to explore the nuances. Somehow greatness seems to be poised between absolute determinism and absolute freedom.

With the preceding in mind, we can see more clearly the fine line that separates genius from madness. An epsilon optimal for creativity in one field of human endeavor may be disastrously large for pursuing the everyday choices of life. A creative artist may require a very large epsilon to create a new art form, but applying such imagination in dealing with the grocer or banker may cause endless complications. Different fields indeed require different types and degrees of creativity. A creative surgeon may devise a new procedure, but it cannot be arbitrarily different from existing procedures, since they are all constrained by anatomy, physiology, and the existing range of instrumentation.

Ralph Waldo Emerson warned us of the dangers of an overly small epsilon when he wrote "A foolish consistency is the hobgoblin of little minds, adored by little statesmen and philosophers and divines." Alexander Pope, on the other hand, described some of the problems of an overly large epsilon when he penned the lines

> *Thus good or bad to one extreme betray*
> *The unbalanced mind, and snatch the man away;*
> *For virtue's self may too much zeal be had;*
> *The worst of madmen is a saint run mad.*

Creativity in all fields, nevertheless, remains the ability to question the obvious tactical choice of the existing paradigms and to explore new ways of approach. Such behavior is an analogue of the chess program being freed of fixed decisions and allowed to explore at random in some domain. Given educational programs also tend to have built-in epsilons, from a very small value for the rigid rote-learning schools to a very large value for very permissive schools. It would be helpful to know a little more about optimal values of epsilon, and indeed many arguments about education center on varying intuitions as to the best epsilon value to try to achieve. Until we know more about optimal values there is always the fun of classifying friends as high-epsilon, medium-epsilon, and low-epsilon types.

Good News for Humanists

Really deep concepts seem to take about 50 years to sink into the collective conscience of the thinking community. So it is that only now are most of us beginning to sense the full impact of certain ideas that have been brewing in physics since the first quarter of this century. For humanists there is good news in this emerging philosophy of science; it says that mankind counts. The world is not something remote and alien to us, but is in some mysterious and ill-understood way the product of an interaction between external events and our creative thrust. This is a message we have been getting from the cultural anthropologists for many years as they studied the perception of the universe by different peoples with totally divergent mind sets. It appears that the viewpoint that is evolving in physics is even more radical and must ultimately have a profound effect on philosophy and the arts.

During the 19th century, physicists developed a rigid, causal, deterministic view of reality in which the entire state of an evolving system followed from the initial conditions. Under these circumstances physics was very far from biology where predictability was rare, and complexity prevented the exploration of causal laws. Toward the end of the century certain paradoxes arose that caused a transformation in the whole structure of natural science. The change of viewpoint was expressed by Einstein (so many thoughts of our age originated with this remarkable man). He postulated that observations in mechanics and electromagnetism depended on the motion of the observer and further demonstrated how different observers would have different perceptions of the same event, depending on their relative motions. While this at first seemed like a mathematical device for resolving certain discrepancies in classical theory, it went much further in relating the scientist and the system he was studying in a more profound manner than had previously been envisioned.

Next came the rise of quantum mechanics and the uncertainty principle, which showed that the act of measurement alters the system so as to make the state indeterminate. But measurement, in the end, is an interaction of a human being with objects in the world around him, again stressing the profound relationship of the scientist and his phenomenological setting. The kind of problem this raises is best seen in the now classical paradox of "Who killed Schrödinger's cat?" In this problem a cat is placed in a box with a capsule of cyanide that may or may not be released depending on a random event such as detecting a cosmic ray. The probability of the event is made one half by choosing the time interval during which the cat is in the box. Quantum mechanics then represents the system as a linear combination of two functions, the dead cat function and the live cat function, each with equal probability. The selection between the two functions involves an observation, i.e., looking into the box. The paradox of quantum mechanics is whether it is the cyanide that killed the cat or the act of observation. For prior to an observation, with no knowledge of the system by an observer, the cat is 50% alive in terms of the theory. After the observation, the cat is 100% dead in half of the cases. If this seems like nonsense, it only reflects the deep problems that have been encountered in relating the observer to the observed in modern physics.

The last of the major branches of physics to yield to the problem of the observer has been thermodynamics. Originally established in a framework where properties were thought to be unique to the system, thermal physics also had a skeleton in the closet. For while the observer was formally excluded in the theory, he was lurking as a manipulator who moved pistons, slowly changed external conditions, and provided isothermal reservoirs. With the coming of information theory, there dawned the realization that entropy was a measure of the observer's ignorance of the detailed atomic state of the system. Once again the observer had become an essential part of the structure of the formalism.

What emerges from these developments is the necessity of dealing with the fact that science does not make purely abstract statements about the world out there but studies the interactions of humans with their surroundings. Physicists, who once thought that it would be possible to avoid the theory of knowledge, are once again confronted with the observer, the human being as an irreducible part of the system. The hard-nosed facts of the science itself are forcing us to a more abstract philosophy where physical reality is some kind of interplay between people and events. A more metaphysical outlook is settling on the subject.

While these changes of viewpoint were going on in physics, biology had, in a certain sense, been moving in the opposite direction. Molecular biology developed from an often unstated philosophical position that its aim was to reduce the observed phenomena in the living world to explanation in terms of the laws of physics and chemistry. These laws, for the molecular biologist, were causal, deterministic, and independent of the observer. Thus the philosophical assumptions of biology were those of physics 50 years earlier. The biologists often tacitly assumed that we could progress from atoms to molecules to organelles to neurons to mind. While the physicist was coming to feel that his view of reality was in some complex way linked to a mind that was outside of his theory, the biologist was coming to view mind as a natural consequence of an established physical reality. The mind-body problem is clearly back with us and is as enigmatic as ever. We are being forced to rethink positions that go back to Descartes, Newton, Hume, Kant, and other earlier thinkers.

Prior to the development of our age, the mind-body paradox involved either assuming all is matter or all is mind or grappling with a dualism that left unresolved how mind and matter could interact. The role of the observer in physical measurement means that the accessible properties of matter now depend on the human mind. In a flashback to some ancient Greek philosopher we once again note that man is the measure of all things.

To the humanists who have worried that science was creating a world apart from man, these results must provide a measure of relief. If we are philosophically no clearer about the ultimate nature of reality, at least the picture is not independent of what we think.

Return of the Six Million Dollar Man

Earlier, I discussed the high cost of biochemical constituents in the human body, starting with the atomic components and going through the hierarchical levels to the incalculable value of man. Since its original publication, this article has been widely reported on and reprinted in diverse publications around the world. One day the mail brought a letter from a childhood friend who is married to a practicing minister. She wrote to tell us of having seen an abstract of the original essay in a church bulletin and requested more information. There were numerous other notices and comments on this estimation of man occurring in a wide variety of religious publications.

All of this has set me to thinking about what it is regarding the cost of human chemical components that is so very interesting to people concerned with spiritual values. Why should it be of concern to the theologians and practicing clergymen that a scientist regards people as materially precious from the point of view of the molecular information richness of the constituent parts? After all, the Judeo-Christian tradition has always recognized that humans are on the one hand "dust" and on the other hand of infinite worth. The significance of a human life, indeed of animal life, is a constant Biblical theme that would hardly seem to gain from biochemical reinforcement.

Western society has, however, been remarkably schizo-phrenic on this issue of human life. Most formal theological doctrines have stated that individual existence is so precious that murder, suicide, and abortion are "sins." The present-day abortion controversy is a clash between that principle and other more recent social goals. Simultaneously, the young of almost every generation have been sent off to perish in war, while the gallows, guillotines, firing squads, gas chambers, and electric chairs have been busy establishing social justice. The degree of contradiction that we cheerfully accept is, after all, one of the enigmatic properties of civilizations.

The theological enthusiasm for the high price of biochem-icals is still worth pondering. Somewhere along the way, starting with the industrial revolution, our culture has become so overwhelmed by technology that we sometimes forget to stand back and proclaim "What a wonderful creature is man." In spite of all dogma and belief to the contrary, the theologian along with the rest of us gets caught up in the machinery, mass communication, and digital printouts, and deep down, I suspect, he begins to wonder about the intrinsic value of the individual. While proclaiming the spiritual worth of each man, woman, and child, he cannot blind himself to the realities of mass culture, loss of individuality, and attendant dehumanization. Therefore when someone emerges from the laboratory, the temple of technology, and says, "Look! Homo sapiens is a very, very precious entity," people from all walks of life are somehow comforted.

If that is indeed the case, then clearly the obligation of the scientist is to step away from the laboratory bench on appropriate occasions and spread the word. Independent of one's sectarian beliefs, there stand the awesome realities of the human body and the human mind. We have only begun to fathom the intricacies of cellular and physiological control mechanisms. When we turn to the nervous system we do not even know the logic, the underlying mathematics, by which the neural network operates. On the mind-body problem we presently lack even the epistemological background to formulate questions in ways that are generally satisfactory. When it comes to knowledge of ourselves, our present limitations remind one of Melville's words, "The whole book is but a draft, nay, but the draft of a draft. Oh, Time, Strength, Cash and Patience." *Awe* has the salutory effect of inducing that precious human virtue, *humility*.

My present concern with religious thought has revived an idea that has been rattling around for a while in the recesses of the cerebrum. Most biological and medical scientists I know are either personal agnostics or somewhat reluctant to discuss theological matters with their colleagues. If they spend their Sunday or Saturday mornings in worship they do not make a point of discussing that fact during the week, at least not with their co-workers. There have been life scientists such as Edmund Sinnott (*Biology of the Spirit*) and Pierre Teilhard de Chardin (*Phenomenon of Man*) who have tried to draw theistic conclusions from biological knowledge, but the average contemporary biologist regards himself more as an heir to the tradition of Thomas Huxley fighting those divines who would deny the reality of evolution and other cherished views.

However, I have the feeling that the most hard-nosed reductionist must be a bit of a concealed theologian when faced with the wonders of the human machine. Most thinking scientists are in fact deeply impressed by the structures that pass through their hands. The unity of the metabolic chart tying all of biochemistry together is truly inspirational. The universal role of ATP and the universality of the genetic code are two powerful examples of the oneness of nature. The facts of bioenergetics, immunology, and morphogenesis are luminous as well as illuminating. It is hard to believe that there is a single researcher who does not undergo occasional episodes of awe and wonderment.

Thus we begin to see two sides of a complex issue. On the one hand are the religious, whose public certainty may be accompanied by a tinge of envy of the scientist who can establish "facts" and test theories by experiment and observation. While such an individual may feel at one level of abstraction as if his knowledge is revealed and complete, he nevertheless looks over his shoulder to see what the scientist is doing and is cheered by the biological message, "How wonderful is man." On the other hand we find scientists who occasionally in the privacy of their offices puzzle over the strangeness of atoms and random thermal energy leading to such wondrous results. The scientist as closet theist and the theologian as closet doubter are both displaying their humanity in a most characteristic way, for uncertainty is a perpetual aspect of the human condition. At times, alone with their thoughts, they may not be as far apart as they think. For my money they're both six million dollar people.

PART II

Christ, Clausius, and Corrosion—SOME THERMAL UNDERTHOUGHTS

Christ, Clausius, and Corrosion

Surely the number of historians of science who have credited Jesus of Nazareth with being the discoverer of the second law of thermodynamics is very limited. Yet when we turn to verse 6:19 of the Gospel according to Matthew we find the following well-known words: "Lay not up for yourselves treasures upon earth, where moth and rust doth corrupt, and where thieves break through and steal: . . ." In the middle of the Sermon on the Mount we discover, to our surprise, that 1,800 years before Rudolf Clausius began his studies, it was already clear that the disorder of the universe had a maximizing tendency.

Very early in the history of man it must have been quite obvious that things left standing deteriorate. The observation was, no doubt, first made for biological objects, since in addition to moths there are rats, cockroaches, termites, fungi, bacteria, and countless other species that are eager to cycle our biological treasures. At that stage of human history, the concept of treasures may not have been a very important one, since life was a very hand-to-mouth affair. There are still primitive cultures, such as the Siriono of eastern Bolivia, where food is rarely kept for more than a day. When the rise of agriculture led to a settled existence, storage and preservation of food became a more serious problem.

43

The concept of a treasure, an individually owned "permanent" possession, could not have been widespread until the late Stone Age or early Bronze Age. With the rise of manufacturing, objects of value and more or less permanence came into existence. However, manufacturing and metallurgy produced products that were permanent only in a limited sense. Chipped stones eventually lost their sharp edges and metallic surfaces oxidized. The problem of rust entered history with the rise of the Iron Age and has been with us ever since.

By the time Jesus of Nazareth stood on the mountain and preached to the assembled multitudes, an empirical generalization was in the air: all things were subject to rot and decay, and there was no eternal permanence to the things that people treasured. Perhaps the time of Jesus is too late in history to date this empirical law. It was clearly implicit in the earlier Buddhism and the doctrine of refusing to consider anything whatsoever as stable and permanent. One is reminded of the dying words of Buddha: "Behold the body of Tathagata. All compound things are subject to decay. Work out your end with diligence" (*Buddha and Buddhism*, Maurice Percheron, Longmans, Green and Co., Ltd., London, 1957). Five hundred years later, the Sermon of Jesus strongly reaffirms in its terse manner a deep understanding of the instability of nonequilibrium structures.

The impermanence of material objects provided no problem for Buddha, who preached a doctrine of rejection, or for Jesus, who advised his followers to ". . . lay up for yourselves treasures in heaven, where neither moth nor rust doth corrupt." However, when we look at the world 1,800 years after Jesus, at the time of the scientific formulation of the second law of thermodynamics, the perspective that prevailed in Western Europe and America was very different from the eschatological views of the followers of Jesus, who believed that the end of days was very near indeed. In the mid-1800s, the industrial revolution was in full swing and an optimism about the future prevailed. Perhaps the most strident voice of this optimism was that of Herbert Spencer, whose early writings envisioned evolution as moving the world to higher and higher states of perfection.

In this expansive era of the industrial revolution, thermodynamics and the efficiency of steam engines was a very important subject for physicists, although it is a standing joke among thermodynamicists that the steam engine did more for thermodynamics than thermodynamics ever did for the steam engine. As part of this study, Clausius introduced the empirical generalization that heat cannot, of itself, pass from a colder body to a hotter body. This innocent-sounding statement became the second law of thermodynamics and led Clausius, by a very tight chain of reasoning, to the conclusion that the universe is tending toward a state of thermal death in which all natural processes will cease. He introduced the concept of entropy as a measure of the disorder of a system and was forced to conclude, "Die Entropie der Welt strebt einem Maximum zu" (the entropy of the universe tends toward a maximum). Thus, starting from the most materialistic foundations of any science, a concern with the efficiency of steam engines, we find Clausius, as well as Kelvin and other contemporaries, turning toward a metaphysics of radical pessimism comparable to the view of the material world enunciated by Jesus and Buddha. This feeling has tinged all subsequent intellectual efforts and is a major component of our contemporary world view. Thus L. E. Loemaker writing in the *Encyclopedia of Philosophy* (The Macmillan Publishing Co., Inc., New York, 1967) notes: "But the theory of natural selection and the second law of thermodynamics, which has been held to imply an end to the universe at a finite time in the future, have put the issue of the destructiveness of natural powers, animate and inanimate, [moths and rust—author's note] on a more objective basis by casting serious doubts upon the possibility and the goodness of evolution and progress." Twenty-five hundred years took us from the religious realization of the inevitability of death and decay to the scientific acceptance of the same principles.

One hundred years after Clausius, 1,900 years after the Sermon on the Mount, and 2,500 years after the death of Buddha—and here we are, still fighting rot and decay. The hardware store is full of moth repellents, rust-resistant paints, and locks against thieves. Although many things have changed in the last few millennia, the universal tendency to decay is still alive and well in the world. This is nowhere more clearly seen than in the problem of corrosion, which has an annual cost of several billion dollars in the United States alone.

The most widely known type of corrosion is the oxidation of metals. Very few of the commercially used metals are found in their elemental forms on the earth's surface. Exceptions are gold and, on occasion, silver and copper. Generally, the ores consist of oxides, sulfides, chlorides, or other combined forms of the metal. The process of refining involves reducing an ore to yield an element in a chemically uncombined form. Such processing requires extensive energy input to break the strong chemical bonds between the metals and the oxidants. The metals are then used to make our treasures, such as automobiles and scalpels. However, the uncombined forms are usually unstable and subject to attack by atmospheric oxygen, and they continually and inexorably degrade to the thermodynamically stable oxidized form. The treasures must be replaced, and we pay a continual maintenance price—our unending debt to the second law of thermodynamics.

The next most widely considered form of corrosion is called aqueous corrosion and involves electrochemical attack on metallic surfaces that are immersed in water solutions. Complex circuits are set up between unlike metals, and the flow of electrons through external circuits leads to the solubilization of metallic parts. It may occur in a household plumbing system, an implanted metallic prosthetic device, or an industrial liquid flow system. The result is always the same—the metallic part eventually has to be resurfaced or replaced.

An entire branch of engineering has therefore grown up to develop methods of combating corrosion. This is obviously of great economic importance as a noticeable fraction of the gross national product goes into corrosion replacement each year. We can get considerably more clever and more active in preventing rust. In the end, however, we cannot win the battle—we can only lose more slowly. For, as the early religious leaders made abundantly clear, nothing material is permanent. If our reason is now thermodynamic rather than theological, the take-home lesson is the same. Strange how such widely divergent branches of human thought should converge on the same idea. It sort of makes one think. . . .

The Entropy Crisis

Each age seems to receive special designation indicating some scholars' views of the distinguishing features of that age. We appear to be entering the Age of Crises, marked thus far by the energy crisis and the environmental crisis. I would like to nominate a new candidate, the entropy crisis. Since entropy is a physical scientist's measure of disorder, there are those who might go further and call the whole age the Age of Entropy.

An understanding of our present problem goes back over one hundred years to Rudolf Clausius, who enunciated the dictum that "The Energy of the Universe is constant; the Entropy of the Universe goes to a maximum." Our modern "local" formulation of the second law of thermodynamics recognizes that every process going on in the material world has a built-in tendency toward molecular disorganization. This disorganization or mixed-upedness is quantitatively measured by the entropy. From an alternative point of view it has been shown that the entropy measures our lack of information about the detailed molecular configuration of a system, such lack being caused by the universal randomizing tendency.

If the entropic factors just discussed were the only principles operative in the world, the surface of our planet would long ago have decayed into a homogenized uninteresting uniform state. Fortunately, from our point of view there is another principle, less well understood, which indicates that by doing work one can counter the disorganization of the second law and cause the organization of systems. Nevertheless, the second law is always there, so that it is insufficient to just order a system and then hope that it will remain ordered. It is necessary to constantly perform work to maintain organization.

The organization that presently exists on the surface of the earth is primarily of two types, geological and biological. The work necessary to develop these two types of order has come from the original gravitational energy of our stellar system, either directly in the formation of the planet or indirectly in the flow of solar radiation. The geophysical processes, such as volcanic flow or metamorphic formation of rocks, lead to the production of distinct types of crust varying in chemical composition. The biological processes are driven by photosynthesis and have led to coral reefs, limestone deposits, sulfur deposits, fossil fuels, and a wide variety of other structures. These phenomena have been going on for a long time and have led to the organization we presently see on our planet. For all of these eons the processes were such that the work-driven organization always predominated slightly over the entropy factors; so the earth evolved from a less ordered to a more ordered state.

The entropy of the universe still increased with the dissipation of the sun's energy and the gravitational potential of the contracting planet. But from this universal tendency toward disorder our planet managed to extract order and to evolve to higher degrees of organization.

The fact that organization in biology is basically rooted in entropic factors was pointed out in a very foresighted way by Ludwig Boltzmann as early as 1866. He wrote:

> The general struggle of the living beings for existence is therefore not a struggle for materials nor for energy (that is

present in every body and in large quantity as heat, unfortunately not interchangeable) but a struggle for the entropy that becomes available in the transition of the energy from the hot sun to the cold earth. To exploit this transition as much as possible the plants spread out the immeasurable areas of their leaves and force the solar energy in an as yet unexplored way to carry out chemical syntheses of which we have no idea in our laboratories.

What Boltzmann called the struggle for entropy we would now read as the struggle for organization against entropic tendencies. Some 80 years later another physicist, Erwin Schrödinger, reiterated Boltzmann's point with the poetic phrase that organisms live by eating negentropy. What was implied was that organisms achieve order against entropic forces by doing work made possible by the flow-through of solar energy. This local struggle against entropy has been taking place on the surface of this planet for over four billion years.

Beginning with the industrial revolution, mankind has been on an entropy rampage, mixing up precious resources that have been sorted out over the long distant past. Carefully sequestered fossil fuels such as coal and oil have been burned back to carbon dioxide and water and mixed up with the atmospheres and seas. Metal oxides have been taken out of veins, reduced to pure metals (an entropy decrease) that are then strewn over the surface of the earth and reoxidized back to their initial state. Entropy from burning the fuels to reduce the oxides has been added to the entropy of dispersing the metals from the high concentration areas in which they were found.

Sulfur has been moved from deposits laid down by countless billions of bacteria or from areas where it has crystalized out of volcanic melts. These pure crystals have been incorporated into hundreds of industrially important compounds, which have ended up in the rivers, oceans, and atmosphere. Once again a mixing up of organized components leads to an entropy increase.

The law of conservation of matter assures us that there is no net consumption of the precious atoms of all the rare chemicals that we require. However, thermodynamics assures us that the more widely materials are dispersed the greater the energy that must be expended to recover them. High-entropy systems require great amounts of work to restore them to an ordered state.

The energy crisis will ultimately be resolved. Either controlled fusion or efficient ways of harvesting solar energy will one day provide abundant power for the earth's needs. The entropy crisis will, however, remain; the mixed up condition of the world's resources will make efficient recovery extremely difficult, if not impossible. The energy crisis is real enough, but it will pass, leaving the entropy crisis as a continuing struggle of future generations.

Obesity: The Erg To Dyne

It seems strange, in view of the enormous literature, both popular and technical, on obesity, that there has been little effort to relate this problem to its basically thermodynamic foundation. The conservation of matter, conservation of energy, and universal tendency toward disorder are some of the ultimate governing laws that determine whether an individual experiences joy or sorrow when stepping onto the bathroom scale. Although the problem of weight control is greatly complicated by the sophistication of physiological feedback mechanisms, a thermodynamic view can establish some limiting relations.

Surrounding each fat man or woman, we may consider that there is an imaginary surface that conceptually serves to separate the person or thermodynamic system from the surroundings that thermal physicists have expansively designated as the rest of the universe. Across this surface, matter and energy flow, and the job of the thermodynamicist is to analyze these fluxes and relate them to changes in the system. He is a bookkeeper of mass and energy who must assess the consequences of these assets and liabilities.

In classical or equilibrium thermodynamics a complete knowledge of the bookkeeping totally defines the system in all of its measurable properties. Living organisms are, of

52

course, the example par excellence of nonequilibrium, indeed of far-from-equilibrium, systems. Nonequilibrium theory is in its formative stages and is incompletely developed. The power of the method is thus somewhat limited and allows less than full understanding and control of a given problem. Nevertheless, certain insights can be obtained.

The law of the conservation of mass, applied to the problem at hand, states that change in the mass of the system is equal to the sum total of flows across the surrounding surface. The major mass inflow is ingested food and the major outflows are respired gases, feces, urine, and evaporated surface water. In the most simple-minded view, controlling weight consists of minimizing the inflows and maximizing the outflows. This view does not take into account the essential energetic aspects of the problem.

The law of the conservation of energy, applied to human subjects, states that the change of energy content of the system is equal to the total of the energy flows across the bounding surface. The principal energy inflow is the chemical energy of the ingested food, while the principal outflows are heat, external work, heat of vaporization of water given off in the vapor phase, and chemical energy of excreted matter. Thus the mass flows and energy flows are completely interrelated, a fact that has long been recognized in assigning an energy unit, the calorie, to the evaluation of food.

Consider in more detail what happens to the energy of the ingested food, more properly the chemical energy of the food and inspired oxygen relative to their thermodynamic ground state of carbon dioxide and water. The energy is 1) used to maintain the organism in a far-from-equilibrium state, 2) used for external work, 3) stored largely as hydrocarbon, 4) excreted as fecal material, which still possesses a high chemical potential.

The first use of energy listed above is to counter the effects of the second law of thermodynamics. A human being is a far-from-equilibrium object and there exists in nature a universal disordering tendency that breaks down ordered structures and drives them toward equilibrium, the state of maximum disorder. To counter these disordering tendencies we must do work to rebuild the ordered structures, and this involves the expenditure of energy. This component of the human energy budget is measured by basal metabolism and presumably represents the minimum energy needed to keep the organism from decaying. It is this battle of ordered systems with the entropic tendency that guarantees us that if we stop eating, our mass will decrease. This is because the carbon dioxide is produced and expired when stored fuels are converted into the energy necessary to maintain the molecular organization of the system, that is, to maintain life itself. In the evolutionary achievement of a very responsive and hence a very competitive system we have come close to the brink of entropic disaster and require a constant supply of fuel and oxygen to overcome the breakdown. Since oxygen is not stored, it is the more stringent requirement.

Another feature of maintenance is the stabilization of the body's temperature at a constant value, which is usually in excess of that of the surroundings. Since spontaneous flow of heat is from a hotter to a colder body, there is a steady loss of heat energy from the system to the surroundings. This heat is produced by metabolic processes that lead to a net flow of carbon dioxide and water out of the system.

The heat produced depends on the volume of an object, hence on the cube of the linear dimensions, while the heat loss depends on the area or square of the linear dimensions. Therefore, large-volume individuals lose a smaller fraction of their total energy as heat. Thus, for example, small mammals must eat an amount of food larger in comparison to body weight than large mammals. This bit of thermal transfer analysis is encouraging to weight losers, since the thinner one gets the higher the fraction of food intake that will go into heat loss.

A fraction of the energy of ingested food is utilized for external mechanical work in walking, running, lifting, climbing, and similar activities. Thus muscles are molecular machines that convert the chemical potential of ingested materials into mechanical work. The detailed nature of the process is not completely understood, but the thermodynamic balance sheet has been studied in detail.

A portion of the energy of ingested food is excreted as chemical energy of feces, material that has approximately the same energy per unit mass of dry weight as the ingested fuel. An example of this energy content can be found by noting that in some parts of the world dry animal dung is used as fuel. The remainder of the energy intake is stored largely as compounds that are predominantly carbon and hydrogen. These fats have a very high energy-to-weight ratio, nine kilocalories per gram versus four kilocalories per gram for carbohydrates, so that once deposited, a large energy expenditure is required per unit weight loss.

It seems ecologically inefficient that an appreciable amount of the energy input of organisms is excreted, but the phenomenon is widespread. The variability of the amount of fecal material accounts for the fact that a small reduction in average food intake may not lead to any weight loss, but simply to a reduction of excreted matter. The utilization ratios of ingested food could become of great importance during a generalized food shortage.

If the net energy ingestion is greater than the maintenance energy plus work energy plus excreted energy, then the system will gain energy. Since this energy is in the form of chemical potential of molecules, in gaining energy the system will gain mass. Similarly, if the ingested energy is less than the sum of the energy loss, weight loss will result. These simple bookkeeping rules constitute the essential thermodynamic guidelines of the problem.

The thermodynamic controls at this point are quite straightforward. Ingestion is an obvious control point, as is external work or exercise. Since we are, however, dealing with people, the thermodynamically obvious is not always within the behavioral domain. Heat loss can be controlled by outside temperature and insulating properties of clothing, but this control may involve some discomfort. The thermodynamic prescription for losing weight, which is to stop eating and swim vigorously in icy cold water, is hardly likely to gain the popularity of Weight Watching or any of the numerous other popular approaches. Nevertheless, the conventional wisdom of weight control is deeply rooted in the precepts of thermal science.

One factor not under direct thermodynamic control is the partition between excreted energy and stored energy. This is a physiological variable rather than a thermodynamic one. For this reason it is the point of attack of all weight control programs that do not involve restriction of total intake. Such programs regulate the quality of ingested material rather than the quantity in an effort to control the partitioning of energy.

The physiology of energy control within the body is a very complex subject, far more complex than the thermodynamic balances we have been discussing. At this point the thermodynamicist must bow out and leave the problem to specialists in metabolism, biochemists, physicians, and nutritionists. The sophisticated science of thermodynamics only confirms the obvious way to lose weight, which is to eat less and exercise more. Any short cuts to this Spartan regime must come from other branches of study. These approaches must involve interfering either with the normal homeostatic mechanisms or with the efficiency of conversion from the potential energy of ingested food to the potential of that ubiquitous energy transfer molecule, ATP.

Let Free Energy Ring

(Written in January 1976
to Celebrate the
United States Bicentennial)

Well, the year of the bicentennial has finally arrived and the spirit undoubtedly is in the air. Indeed, it should be rehabilitating to think back to those founders who dared to "hold these truths to be self-evident, that all men are created equal." I copy these words from a tiny black book called *I Am An American*, which sits on my desk. The volume curiously bears on the title page the signatures of my grade-school principal and my eighth-grade history teacher. It was, I believe, a gift to our graduating class. All of this will probably sound incredibly maudlin to today's thirteen-year-olds, but let's remember that nostalgia is the stuff that bicentennials are made of. Therefore, I must confess that although eager to celebrate the birthday with you, I am reluctant to add my few words to the torrents of prose that will flow this year in praise of "The Unanimous Declaration of the Thirteen United States of America." Instead, I ask you to go back with me 100 years to the second most significant document produced in the United States, Josiah Willard Gibbs' paper "On the Equilibrium of Heterogeneous Substances" (*Transactions of the Connecticut Academy*, III, pp 108–248, Oct. 1875–May 1876, and pp 343–524, May 1877–July 1878). It seems appropriate to commemorate the bicentennial with a look back to the centennial of the most far-reaching scientific paper of American origin. One hundred years ago, the mind of America was fermenting and establishing standards of scientific productivity and excellence that even today must stand as an example to inspire the workers in this field.

If the history buffs of the summer of 1976 venture to New Haven, they may be drawn to the Grove Street Cemetery, where among the monuments of such American notables as Noah Webster, Roger Sherman, and Eli Whitney, one will find a stone bearing the inscription, "Josiah Willard Gibbs, Born Feb. 11, 1839, Died April 28, 1903." On the other side of the stone are the words "Professor of Mathematical Physics in Yale University 1871–1903." Gibbs was, by our standards, an extraordinarily provincial man. Aside from three years that he spent in Europe attending lectures from 1866 to 1869, almost his entire life was spent within a mile of the Grove Street Cemetery. He was born in New Haven, educated at Hopkins Grammar School and Yale, and later spent his entire professional career on the Yale faculty. Gibbs was something of an introvert; he never married and lived with his sister and brother-in-law. Though his day-to-day life was regulated and routine, his intellectual life was a constant pioneering expedition through completely uncharted areas. He was, according to contemporaries, a serene, friendly man whose outward behavior gave no clue to his fantastic scientific accomplishments.

His work in thermodynamics began late in 1871 and terminated in 1878. Although he published in the *Transactions of the Connecticut Academy*, a journal of limited circulation, Gibbs clearly perceived the importance of the work and sent reprints to 90 of the leading scientists of his day. He was neither modest nor immodest; he simply knew the significance of what he had done.

Turning from the man to his work, the paper begins with the famous quotation from Clausius.

Die Energie der Welt ist konstant
Die Entropie der Welt strebt einem Maximum zu.

It was a characteristic of the simpler world of the last half of the 19th century that permitted physicists to make axiomatic statements about the energy and entropy of the universe. Gibbs took these two postulates of Clausius and worked out the detailed consequences that set the direction of physical chemistry for the next 100 years. He introduced the concept that we now call the chemical potential, the incremental amount of energy added to the system when a small increment of a chemical component is added. This construct, when combined with the existing equations of thermodynamics, allowed Gibbs to deduce most of the important relations of chemical equilibrium as well as a theory of osmotic pressure. He proceeded from chemical potentials to the phase rule that has become a cornerstone of geological theory and metallurgy. Today, we designate Gibbs' Free Energy as the mathematical functions that he introduced to deal with chemical problems.

The next section of the long paper came to fruition 80 years later in providing the methodology for answering one of the most penetrating questions in 20th century biology. The part of Gibbs' paper under consideration bears the ponderous title "The Conditions of Equilibrium for Heterogeneous Masses under the Influence of Gravity." In the discussion, Gibbs derives the equations for the distribution of chemical components in a system under the influence of gravity. When many years later Theodor Svedberg developed the ultracentrifuge for the separation of large molecules, he realized the generality of the equation of Gibbs. By replacing the gravitational acceleration by the acceleration in a centrifugal field he was able to derive the equations of sedimentation equilibrium. K. O. Pedersen, a coworker of Svedberg's, utilized these concepts to develop centrifugally stabilized cesium chloride density gradients. Pedersen acknowledges Gibbs as the source of the basic equations (T. Svedberg, K. O. Pedersen, *The Ultracentrifuge*, 1940).

In 1957 a group of workers at California Institute of Technology developed cesium chloride density gradients for use in separating macromolecules on the basis of density that could be controlled by isotopic substitution (M. Meselson, F. W. Stahl, J. Vinograd, *Proc Natl Acad Sci USA* 43:581, 1957). The Cal Tech group used, as a starting point, the equation of Pedersen, which is, as noted, directly from the formulation set forth by Gibbs in 1876. The method of gradient separation quickly led to the series of experiments demonstrating the semiconservative replication of bacterial deoxyribonucleic acid. Bacterial genetics as well as the method of gene replication were thus established on a firm physiochemical foundation. Much of the conceptual basis of modern biology and its applications to medicine thus owe a debt to the quiet, serious scholar and his lonely labors.

The rest of Gibbs' long paper reads on like an oracle on the future of science: theory of ideal gases, stress-strain relations in solids, theory of capillarity, adsorption isotherms, stability of surfaces, liquid films, theory of the electrolytic cell, and many others. Physical chemists are still working out the full consequences of what was set down 100 years ago. Gibbs did not stop with this monumental work; he went on to establish statistical mechanics and to contribute substantially to the electromagnetic theory of light as well as to other topics.

Every natural anniversary should have a lesson to teach, otherwise it hardly seems worth celebrating. If the bicentennial is going to teach us the importance of the United States' revolutionary heritage, what is the lesson of the centennial? Gibbs' writings were as revolutionary as the Declaration of Independence, yet creative genius is so elusive we seem to lack the perception to define the conditions for its development. Young Gibbs certainly had the leisure to follow his own inclinations. His first three "postdoctoral" years were spent touring Europe attending lectures he deemed most interesting. He did not publish a paper until seven years after he was awarded his doctoral degree. He never applied for a grant; indeed, he did not even receive a salary for his first few years on the faculty. He was never spoiled by success nor was he so overcome by a desire for recognition that this was a major item in his life. He was himself, of tremendous internal resources, integrity, and self-assurance. Perhaps that is the lesson to be extracted from this centennial. In the squabble for jobs, promotions, school admissions, tenure, grants, and all the superficial decor of success, do we remain true to some inner sense of integrity?

Gibbs was a free man, much as was Jefferson, free because he was at peace with himself. Which of us knows who he is, in spite of what colleagues, chairmen, heads of service, deans, reviewers, and grant administrators may think? Is this not the central theme of mid-19th-century America, seen in men like Gibbs and serenaded by writers like Thoreau and Emerson? If ever an individual possessed "self-reliance," that individual was Josiah Willard Gibbs. Perhaps, therefore, this savant can teach us about two types of equilibria, thermodynamic and human.

Whose Energy?

In one of those rare interactions between academia and the counterculture, I found myself listening to a lecture on the theory of massage, and frankly it rubbed me the wrong way. The irritation came about over the speaker's use of the word *energy*. It became clear, as he spoke of gathering of the body's energy or transfer of that stuff from the masseur to the recipient, that he was talking about a construct that obeyed neither the first nor the second law of thermodynamics. This nonphysical use of *energy* can be widely found in diverse modern writers on yoga, Zen, massage, meditation, and various metaphysical schools of thought. The freedom in defining—or failing to define—terms is a cause of some concern, since the problem of energy supply has become such a pervasive worldwide technological and political issue. Particularly upsetting in the talk on massage was the fact that the bioenergetic notions of the speaker were close enough to the ideas of normalized science to imply that they were the same, even though the masseur's statements were unconstrained by the most elementary laws of physics. This freewheeling thought, using the vocabulary of science, is a characteristic of many schools ranging from astrology to branches of psychiatry and, I believe, universally leads to a kind of mental fogginess.

It seems worthwhile to take *energy* as a case study in semantic confusion and the concomitant actions that are produced. The word itself goes back at least to Aristotle's writings and seems to have implied vigor of expression, impressiveness, or exercise of activity of any kind. The contemporary form entered English through the Latin and acquired additional usages such as strenuous exertion, display of power or potential power.

With the rise of modern physics the concept of energy became a dominant theme. As early as 1807 Thomas Young had written, "The term energy may be applied with great propriety to the product of the mass or weight of a body, into the square of the number expressing its velocity." We would now use the term *kinetic energy* for the measure defined above (with a numerical coefficient of one-half). The next 50 years witnessed the remarkable rise of thermodynamics culminating with the law of conservation of energy as a cornerstone of all physics. Through the efforts of such scientists as Helmholtz, Mach, and Poincaré, energy became a precise metric quantity capable of about as clear a definition as has ever been given to a word for an abstraction. This precision was also a requirement of the parallel developments of the Industrial Revolution, which demanded exacting relations between joules, BTUs, and kilowatt hours on one hand and dollars and cents on the other.

During these advances in physics and engineering, *energy* continued to retain and develop all of its other meanings. For example, in 1861 May wrote of "The troublesome energies of Parliament" and in 1887 Lowell spoke of institutions that "had surely some energy for good." Thus we begin the twentieth century with a very precise scientific usage and a wide variety of imprecise meanings.

The next development in the unfolding of our terminology came with the rise of psychoanalysis and the establishment of such notions as libido or sexual energy. The role of this construct has been commented on and criticized by Szasz (*The Myth of Mental Illness*). He describes the analogy as follows: "Traditional psychoanalytic theory, as well as modern psychosomatic theory, is based on the physical model of energy discharge, of which a hydraulic system is an instance. In such a system, a body of water behind a dam, representing potential energy, 'seeks' release, and may be discharged through several pathways. First, through its proper channel into the riverbed into which the water is intended to flow; that is, through 'normal' behavior. Second, through some other route, such as through a leak at one side of the dam; that is, through hysterical conversion. And third, through another route, such as through a leak at another side of the dam; that is, through organ neurosis." In such examples the language of *energy* has permeated the literature of psychoanalysis.

From the domain of orthodox psychiatry the energy terminology passed into the fringe in the writings of Wilhelm Reich. While there is a strong urge to ignore the biophysical nonsense that Reich promulgated, it is not possible to ignore the influence of his writings and the effect they are having on the use of the word *energy*. Reich started with the psychoanalytical use and moved to a kind of vital force, "orgone energy." He then completely jumbled this concept with the scientific meaning of energy and developed the notion of negatively entropic stuff, primordial in time and hierarchy to conventional matter and energy. He kept interweaving the language of physics into a structure totally removed from the route of empirical verification that is required of science. (For a sympathetic view of Reich see *Wilhelm Reich and Orgonomy* by Ola Raknes.) In the process he kept stressing such terms as life energy, psychic energy, emotional energy, bound energy, residual energy, orgone energy, and bioenergy. These terms circulate among the counterculture but pass into a more academic vocabulary, carrying with them the residuum of Reich's madness.

Another *energy* has entered Western thought from classical Chinese philosophy and physiology. It corresponds to a kind of body tone and is perhaps closest to the idea being used in the lecture on massage. This *energy* is found in the vocabulary of acupuncture and traditional Chinese medicine. As it appears in contemporary usage it seems to have neurophysiological correlates in terms of surface potentials and rhythms that emanate from neural networks. It can apparently only be understood in its own philosophical framework and provides considerable difficulty for traditional Western biology.

A still different construct of *energy* has appeared over the last few decades in the posthumous writings of the religious philosopher Pierre Teilhard de Chardin. In his most widely read work, *The Phenomenon of Man*, he introduces another *energy* with the comment, "There is no concept more familiar to us than that of spiritual energy, yet there is none that is more opaque scientifically." He then proceeds to a discussion of the mind-body problem rephrased in the language of spiritual and material energy, which is the measurable content of thermodynamics. The theologian refuses to accept a dualistic world view and argues, "In the last analysis, somehow or other, there must be a single energy operating in the world." But Teilhard de Chardin was a scientist as well as a theologian and recognized the difficulty of his position. His solution remains unsatisfying to the reader and I suspect was unsatisfying to the author, but it does leave us with yet another *energy* to contend with.

The residual confusion of these concepts can be seen in phrases from course descriptions of a counterculture "educational exchange." "Exercise through energy-current release by posture-stretch-relaxation." "Polarity principles enhance understanding of release of energy blocks in Yoga postures." "Awaken your sensitivity to the movement of your energy." "Basic attention to energy awareness, grounding, alignment and Tai Chi as meditation." "An experimental introduction to bioenergetics or neo-Reichian work." "We will learn to sense psychically our own energy flow and start to work with it." "Astrology . . . the distribution of psychic energy within our solar system."

By now the reader is no doubt fully impressed with the semantic difficulties. While everyone is worrying about who owns the world's energy supply we may seem a bit pedantic worrying about who owns the word. Nevertheless, clear thinking demands that we have an agreed-upon set of definitions for our most basic concepts. This clarity is going to be most useful in the solution of any problem. As a first step I suggest we take a thought from *The Greening of America* and define Energy I, Energy II, Energy III, etc., as Charles Reich utilized Consciousness I, Consciousness II, etc.

Energy I is the measurable conserved quantity of thermodynamics. It is the stuff we pay the power companies for, we are having a crisis about, measure in joules, and pay for in dollars.

Energy II is the notion of Aristotle embodying activity and vigor. It can best be illustrated by the sentence, "I could write plays like Shakespeare, if only I had the energy."

Energy III is the vital body tone of traditional Chinese medical philosophy. Its relation to metabolic and neurophysiological constructs is obscure and awaits the full "Meeting of East and West."

Energy IV is the psychiatrists' concept expressed most forcefully as libido, drive, or related terms. It clearly is concerned with Energy II and III.

Energy V is spiritual energy. For adherents of various philosophical theories it is the motivating and organizing factor behind all other energies and is therefore primary. From the point of view of thermodynamics, it is meaningless.

All of this grows very confusing and illustrates the problems of assigning a myriad of meanings to a single word. As a student of Energy I, I wish people in other areas would use different words; it would help me avoid guilt by semantic association.

Women's Lib and the Battle Against Entropy

There was no doubt about it; the flooring on the small outside porch was going to have to be replaced. It was rotting through from the bottom and presented a real hazard. One trip to the lumber yard and there I was, ripping up the rotted planks and thinking naturally enough about the second law of thermodynamics, which was creating so many jobs. There had been the corroded copper pipe in the kitchen, the dirty ceiling in the laundry, and now the side porch. The universe was clearly and unmistakably moving downhill.

A bit of reflection brought me to the sudden realization of how many of life's activities are directly tied up with our unending effort to slow down the increase of disorder in the immediate surroundings. Those most closely associated with our individual survival are reflected in the physiological concept of basal metabolism, which is a measure of the energy the body spends in maintaining itself far from equilibrium in spite of nature's tendencies toward that end. Thus we must maintain concentration differences that are opposed by diffusion and electromotive force differences that are opposed by leakage currents. Finally we must continually rebuild the delicate protein structures that are breaking down under thermal denaturation and autodigestion. All of these things take energy, and much of our agriculture and food-preparing efforts basically go into just keeping ourselves alive and functioning in our battle with the second law of thermodynamics.

By the time I had come to this conclusion, the floor boards had been ripped off and it became painfully clear that some of the two-by-sixes supporting the floor were also going to have to be replaced. While putting creosote on the fresh lumber there was time again to return to this problem of fighting entropy. I turned attention to thinking about how many jobs around the house were negentropic efforts that led nowhere but simply kept us even in the eternal struggle against the disordering tendencies of the universe.

The second law of thermodynamics states that spontaneous processes tend toward a maximum of disorder, and work must be expended to maintain systems away from this undesirable state. And anyone who has watched the papers on a desk or the contents of a house go through the randomizing process can only be impressed with the power of this principle of nature. All of life is a creative tension between the work-consuming processing on the one hand and the accompanying spontaneous decay on the other. One is reminded of the myth of Sisyphus who spent his time in hell rolling a big stone up a steep hill only to have it roll down before it reached the top. Sisyphus represents the work input, and the spontaneous tendency of the stone to roll down the hill is reminiscent of the second law.

When this profound conclusion was reached my thirst became apparent, and a trip to the kitchen sink for a drink of water was in order. On the way back, a trail of sawdust on the kitchen floor again served as a reminder of the disordering tendencies. But then an idea hit with the resounding boom of the cannon in the "1812 Overture." All of housework is a battle against entropy. Every housewife, doomed to repetitively sweeping the floor, washing the dishes, dusting the furniture, and cleaning the clothes, is devoting her life to fighting the second law of thermodynamics. In this context, a global awesome aspect of women's liberation began to stand out.

Our society, up until a few years ago, had decided that the eternal struggle against disorder was woman's work and that the possessors of two X chromosomes were to be consigned Sisyphus-like to the unending task of countering a law of nature. It was a battle that could never be won. There were no triumphs, no victories; the best one could theoretically do was to break even. Men would build unstable structures (as all structures are ultimately unstable), and the task of maintaining these structures against the pervasive and unending decay tendency would be left to women.

In these terms, one senses the cosmic unfairness of women's traditional role in Western society. Remember that Sisyphus was in hell because of our very intuition that constantly laboring to get nowhere is a vision of hell. To make the whole grievous situation seem better than it actually was, we created a myth that being a successful full-time combatter of entropy was a virtue and developed a beatific image of "the good housewife." The intent was, however, not sainthood but servitude. What no one ever uttered was the thermodynamic truism that perfect order requires infinite work, so that the stated goal was physically unattainable. By the fundamental rules, housekeeping was established as a "no win" game. Viewed in this context, the only saving grace of the woman's role was the ultimate triumph against disorder—the creation of a new human being. Motherhood is a satisfying role, but it has been a high price to pay for all the accompanying entrapment by an ethic that, at best, belongs to an earlier age.

By this time I was so engaged in thought that I was in danger of smashing my fingers with the hammer, as the floor boards were pounded into place. How could such a situation have persisted for so long? Sexual dimorphism is a reality, and indeed in many animal species sexual roles are sharply delineated. But we live in a civilized society. The idea of setting women to fight the second law unaided is just gross unfairness. A law of nature is a law of nature no more for the goose than the gander. We need a constitutional amendment guaranteeing equality before the second law of thermodynamics.

At last I felt that it was possible to place the concept of woman's liberation in its proper context. What should be demanded is neither sexual, nor Freudian, nor even political. What is required is that the job of pushing the stone uphill be fairly distributed. In addition, we must settle for a reasonable entropic state and not exhaust ourselves struggling for an unattainable one. There is no escape; the boulder will come down the hill again. We can, however, add joy to the job by sharing the task. Then we will all have some time left over for the more creative and enjoyable aspects of life. Surely this is a minimum goal for a just society.

It was now beginning to get dark and the planks were completely nailed down. All in all, it had been a successful day. I had counteracted a lot of decay and was well on the way toward making a stronger porch than had existed before. I also felt liberated by the realization that we all have to work together. I celebrated by sweeping up sawdust from the kitchen floor, leaving a few motes to symbolize the eternal power of the second law of thermodynamics.

PART III

Bulls, Bears, and Bacteria—LOOKING AT FLORA AND FAUNA

Bulls, Bears, and Bacteria

What do Merrill Lynch, Pierce, Fenner & Smith have to do with swimming bacteria? This unlikely question came into view a few weeks ago when, by chance, I happened on the same day to both read about commodity speculation and attend a lecture on bacterial motility. My reporting a relation between these two has no doubt created a small credibility gap that I must begin to dispel. We turn first to a fascinating series of recent findings about how and why bacteria swim.

The phenomenon of bacterial motion has been known for over 250 years, since Leeuwenhoek first peered through his microscope and saw his "wretched beasties." Later with improved optics it was found that the swimming microbes have long, thin flagella and are driven through the water by the motion of these organelles. Present-day techniques of dark-field and electron microscopy have provided much more detailed knowledge. It has been established that the flagella are attached to the cell at a basal body, and the turning of these organelles of locomotion provides the thrust for the tiny swimmers. Thus are seen the world's earliest rotary engines, the exact mode of whose operation is still unknown. The molecular details of how chemical energy in the bacterial cell is converted into the mechanical energy of the at-

73

tached rotators is a current challenge to biologists.

It had long been noticed that a swimming bacterium tends to move in a straight line for a while and periodically stops and goes off again in another randomly chosen direction. This change of bearing is known as a tumble and appears to be quite a universal activity for swimming microbes. All of these observations were interesting curiosities until recent findings that have given some functional insight into why bacteria move. The answer comes from the now well-established generalization that bacteria travel toward higher concentrations of attractant molecules and away from repellents. The attractants tend to be nutrients and the repellents tend to be toxins. Thus motile bacteria exhibit what we would call "behavior." This simple-sounding statement is really quite revolutionary in the world of psychology. For a long time now learning and habituation experiments have been carried out on lower organisms, but it was generally expected that below some branches of the phylogenetic tree "behavior" would cease. Microbes clearly seemed far beneath the psychologist's domain. It is now seen that purposeful activity extends to the most primitive living cells bordering on the size range of the molecular.

Additional results in recent years tell us in a more detailed manner just how bacteria manage to swim toward favorable environments and away from harmful ones. By processes as yet incompletely understood, bacteria sense the time rate of change of concentration of molecules in their environment. They then respond by determining the length of time between tumbles. Thus if a bacterium is swimming toward a higher concentration of food it waits long periods between changes of direction, while if it finds it is swimming away from food it quickly tumbles. This results in the net motion of a given cell toward the highest nutrient values. Similarly, if it is swimming toward repellents it tumbles frequently, while if it is swimming away from repellents it waits before changing course. These tiny organisms have thus evolved what we must designate as a strategy to deal with the problems of living in a variable environment.

The introduction of the concept of strategy takes us from the survival struggle at the microscopic level to the equally intense struggle waged by the bulls and the bears. For, as noted, the day I attended the seminar on bacteria I also read *Handbook for Commodity Speculators* put out by Merrill Lynch, and on page three I was surprised to find the statement, "This is just another way of saying, 'Cut your losses short and let your profits run,' which is a basic principle in any kind of successful speculation." The strategy proposed for speculators is the same as that used by bacteria for countless eons. At first this seemed very odd, but a little pondering brought out some interesting points. Both bacterium and speculator live in a variable environment where it is very difficult to predict the future. Both must be prepared to change course quickly. In this situation the best one can do is to monitor events and rapidly respond accordingly. The parallelism in the techniques results from the similarities of the situations imposed by the great uncertainties in foreseeing events. Thus "cut your losses short" means tumble quickly, while "let your profits run" means stay on course if things are going favorably.

The fact that the most primitive of creatures and the most sophisticated capitalists employ the same devices for survival might be taken by some as reflecting on the cerebral attributes of the speculators. I for one take the opposite point of view and think that this finding should be celebrated as a paean to the wisdom of nature. From the very beginning of the evolutionary process, behavioral survival programs were incorporated into the repertoire of actions of organisms. Intelligence is not the sole property of humans; it permeates nature, and we almost begin to see it emerging as a primitive property of matter along with mass and charge.

For some time now we have realized that even lower invertebrates must be able to sense their environments and to respond with behavioral modifications. It is clear that an organism that has this ability will have an advantage over one that does not. In 1906, H. S. Jennings wrote a fascinating book entitled *Behavior of Lower Organisms*. He described a number of stimulus-response experiments illustrative of the point just made. One of these dealt with the single-celled protozoan *Stentor*. Jennings' elegant description of the response of this tiny animal to a cloud of carmine grains introduced into its environment makes it very clear that this organism has developed an elaborate behavioral pattern to deal with an adverse environment. Since this classic work a large number of other studies have confirmed that the protozoa do indeed show a remarkable range of responses.

However, in moving from single-celled animals to bacteria we move across a vast evolutionary gap to the much simpler prokaryotes. If these organisms sense their surroundings and respond with behavioral strategies, how far down in the domain of the molecular must we go to find the first beginnings of intelligence? Perhaps you don't think that these microbes are showing intelligence. Check it out with your own broker.

Loch Ness to Lahaina

It's a long journey from Loch Ness, Scotland, to Lahaina, Hawaii. Yet, somehow, sitting in the bar of the Pioneer Inn in that Polynesian port I could not get my mind off the distant Scottish monster. I had been exercised over people's attitude toward Nessie for many months now, indeed ever since *The New York Times*, that citadel of journalistic respectability, had underwritten an expedition to search for the mythical beast. What had triggered my thoughts about the lake-dwelling giant was talk of actual giants, the humpback whales now in the waters around Maui.

Seated at the table with us, sipping white wine together, were two young members of an organization called Greenpeace, one of whose goals is to save the whales from extinction. For the third year in a row Greenpeace volunteers are planning to place themselves in small boats between the Soviet and Japanese harpooners and the whales. This is an act of consummate courage, since the harpooners, who use explosive charges on their missiles, have proved to be less than respectful of those defending the whales. A second danger comes from the violent thrashing of wounded sea mammals who, in their death throes, are unable to distinguish friend from foe. The Greenpeace members go out, nevertheless, in an act of passive resistance to place their bodies between the hunters and the hunted.

Saving a species from extinction seems at first sight like a very abstract goal for which to risk life and limb, and as we delved into this "heavy" topic, my errant mind began to drift to thoughts of the Loch Ness monster. What has bothered me about the time and effort spent on searching for this phantom being is that it ignores some very profound biological principles. Individuals do not exist in isolation. The proper ecological unit is not the solitary organism but the species. Organisms do not rise de novo but undergo morphogenesis, utilizing genetic materials contributed by parents. The Loch Ness monster must either be a member of a living group or must be the last survivor of a taxon approaching extinction. If Nessie is a member of a viable species, then there must be a critical number of relatives around. Now, one monster might hide in a lake, but the possibility of a whole population of behemoths evading detection stretches the imagination to the limits. If we are observing the last monster, where are the fossil remains of the forebears? An animal so large could not disappear without leaving traces, yet no single bone or bit of detritus has ever found its way to a paleontologist who has proclaimed it as the remains of a *Monsteria lochnessi*.

Thus *The New York Times* was engaged in the journalistic sin of making biological news in Scotland while the real story lay in the oceans of the world where the populations of some species of giant whales were being reduced below the numbers where animals of this ecological type can maintain their populations. However, the communications media have gotten the message, and Lahaina is very temporarily the film capital of the world as five separate professional film companies are off to sea to seek out and record the activities of our cetacean neighbors.

It is ironic that this city, formerly one of the chief ports of call of the whale killers of the world, is now the port of call of those who would preserve, on film at least, the story of the lives and loves of the giants of the deep. But preserving this material on film, valuable though it may be, is a poor substitute for maintaining the real thing, a self-sustaining population. One wonders if the filmmakers have sensed that the time is late and this may be their last chance.

This is where Greenpeace and other organizations dedicated to saving the whale enter the picture. They are unwilling to concede that the battle is lost, and they venture forth in simple human acts. This is not the work of governments, bureaucracies, or other ponderous human institutions that are unable to cope with so simple a problem. This is the work of individuals who devote their time, their efforts, and perhaps their lives to pursue an ideal, a strangely biological ideal.

Now, as all of us who are familiar with evolutionary history know, species arise, flourish for a while, and become extinct in an unending adaption of life to a changing planet. What is different now is that one species, *Homo sapiens*, alters the planet so rapidly that an orderly evolutionary progression is no longer possible, and massive extinctions are the rule. What is particularly disturbing to the biologist about the whale story is the thought of losing so unique a group of animals. They are special in at least two important ways. Firstly, they include the largest mammals (and the largest animals) that have ever existed. There is thus much to learn about basic physiology, including thermal regulation and the circulatory system. Secondly, and more important, they appear from a number of points of view to be the second most intelligent group of mammals around. Whales, therefore, may indeed have much to teach us if we and they can learn to communicate. Such knowledge will be irretrievably lost with each species that becomes extinct. Besides, if these creatures are indeed so like us in mind, if not in body, then killing them seems an act of fratricide, and eating them seems an act of cannibalism.

The best known whale story to date is *Moby Dick*. It has not only influenced our thinking about whales, but it has had a deep impact on the American literary tradition. But perhaps the greatest story about whales is yet to be written.

While the first epic dealt with the mad captain dedicated to the extinction of the white terror, the second may deal with the seagoing madmen of Greenpeace, dedicated to preserving cetacean life. If only one had the skill of a Melville, perhaps, perhaps? It can get pretty heady around the Pioneer Inn after a bit of wine, and besides, the ghosts of Ahab, Starbuck, Ishmael, Queequeg, and the rest are whistling through the rigging of the boats in the harbor.

Godspeed, Greenpeace!

Of Mammals
Great and Small

Limiting cases often attract our interest; the greatest show on earth needs its giants and midgets. So too in biology, extremes often give us an opportunity to study special aspects of groups of organisms. Man's near relatives, the placental mammals, present some thought-provoking examples of the consequences of size variation.

Consider then these higher organisms. They are members of a biological class of about 3,500 species who are warm blooded, hairy, and nourish their young on the secretions of special glands of the mother. The most numerous, in both variety of species and numbers of individuals, are the rodents. But more than the diverse taxonomic features, mammals share a common physiological and biochemical strategy for dealing with the problems of function and survival. All thermoregulate within a few degrees of the same temperature, and all share a basic body plan in terms of organ systems.

What is fascinating from an evolutionary point of view is how incredibly diverse in outward form and habitat the mammals have become in the relatively limited span of about 80 million years. It is difficult to look at a blue whale, the largest living mammal, and the Etruscan shrew, the smallest living mammal, and to realize that they are at most a few million generations removed from a common ancestor. We must then examine these two organisms, the mighty and the minute, to see the limits to which evolutionary thrust has carried this one class of animals.

Suncus etruscus, known as the pigmy white-toothed shrew, is found along the Mediterranean coast of Europe and some surrounding areas. This tiny, inconspicuous animal is a member of the order Insectivora, a not very well characterized group that includes the shrews, moles, and hedgehogs. They are regarded as the most primitive placental mammals, the closest relatives of the early ancestors that coexisted with the dinosaurs.

The blue whale, *Balaenoptera musculus*, is the largest of all whales, the largest mammal and probably the largest animal that ever existed. Indeed, "prodigies are told of him" (*Moby Dick*, Chapter XXXII). Found mostly in the Antarctic and Southern Hemisphere, blue whales are fast as well as large and can swim at speeds of up to 20 knots.

The first critical comparison between whales and shrews comes when we consider the weights of the two species. Weighing a shrew is relatively easy, but putting a whale on a scale is a matter of some gravity. The whole animal must be carefully cut up, and it is necessary to keep track of all the components, including fluids. The record for a blue whale is a 1950 measurement of 134.25 tons, although there are probably larger specimens (W. C. Winston, *Natural History*, 59:393, 1950). Etruscan shrews seem to average about 2 gm, while adult specimens as small as 1.6 gm are known. Thus, between the upper limits of whaledom and the lower limits of shrewdom, there is a size ratio of 100 million to one. A difference so large within a single relatively recent biological class makes one wonder if our evolutionary ideas are adequate or complete.

In trying to compare our two creatures, we find that much of the information is lacking. Extensive dissections of large whales are hampered by technological constraints. The problem has been described by the Dutch vertebrate anatomist

E. J. Slijper: "The heart of the whale has to be dragged over the slippery deck of a whaling ship by seven strong men—and then with some difficulty. The poor biologist who wishes to cut up this huge mass, almost 6 feet wide and 10-11 cwt, may well shy from such an Augean labour." The Etruscan shrew, on the other hand, seems to be small enough to have been largely overlooked by comparative anatomists with sharpened scalpels. What is clear, however, is that both creatures have the same overall body plan and undergo the same type of morphogenic process from fertilization to birth.

If we examine the dimensions of cells, it seems apparent that the size of a given cell type is almost independent of the size of the animal. Thus, whale red blood cells are slightly larger in diameter than those of humans. However, the horse, intermediate between the two mammals just cited, has erythrocytes even smaller than those of man. The most extensive comparative study of cell size that I have seen involves the dimensions of liver cells in 10 species of mammals varying in weight from the rat (about 100 gm) to the cow (about 1 million gm). Over this range of animal size, the liver cell diameters vary by no more than 20%, and this is uncorrelated with the animals' bulk (W. Pfuhl in *Handbuch der Mikroscopischen, Anatomie des Menchens*, 1932). Only nerve cells appear to vary somewhat systematically with the size of the animal. The variation is relatively minor, so that elephant neurons have about twice the diameter of similar mouse cells. The message from biochemistry and cytology appears to be that a mammalian cell is a mammalian cell, and the great diversity we see is at a higher level of organization. Even grossly, there are some surprising regularities. Thus, the ratio of blood to body weight is 6.5% in blue whales and 5% in rats.

In general, it is thought that the lower limit of mammalian size is imposed by heat losses resulting from the geometric fact that in scaling down, the area decreases much more slowly than the volume. Thus, while a large shrew consumes 5.5 cc of oxygen per gm per hour, a lesser animal has been reported to consume 11 cc per gm per hour. To keep up this rate of metabolism, a tiny shrew has to eat well over its own weight per day in food. If the animal got even smaller, this food ratio would have to get proportionally larger. *Suncus etruscus* exhibits a marvelous adaptation to this constant threat of starvation. If the animal is cold and starving, it goes into a state of lethargy, resembling hibernation, in which its temperature drops to about two degrees above ambient, thus sparing the metabolism. It awakens in a few hours, becomes active, and seeks food for a short time. If none is found, it goes back into the lethargic state and repeats the process. It can thus survive for a day or two during periods of food shortage. Blue whales maintain a considerably higher metabolic rate than might otherwise be expected from their size, since they spend a large portion of their life in arctic waters. The rate of heat transfer to water is much greater than to air, so that such mammals require a large food intake. While there are no metabolic measurements of blue whales, the bottlenose dolphins require about 500 kcal/kg/day.

The upper limit of size on land is the force of gravity, which imposes severe structural requirements on large animals. In the sea, the upper limit must ultimately be ecological in origin. Blue whales live largely on krill, a collection of small crustaceans. As much as one ton of krill has been found in a cetacean stomach. To obtain adequate food, the predator must sweep out large areas of ocean. A sufficiently large animal would eventually reach the point where it could not get enough food to support its metabolism. Thus, curiously enough, at both ends of the size range, mammals are limited by their energy intake.

The scaling of glands among the mammals presents some insights on size. The pituitary of a cow weighs about 2 gm, whereas that of a blue whale weighs about 35 gm. Assuming an approximate constancy in the ratio of pituitary weight to body weight, we would anticipate that for the pygmy shrew the gland would weigh 0.35 μg, and the posterior pituitary lobe, by the same reasoning, would weigh only a fraction as much. However, since mammalian cells weigh on the order of 0.003 μg, the posterior lobe would have only about 100 cells. Here we see another feature of the small size range. The necessary organ complexity and the approximate uniformity of cell size impose an ultimate cellular limitation.

This brief look at the animals at the extreme ends of the size scale makes it evident that we are lacking much of the necessary knowledge to understand fully the mammalian domain, its evolutionary history, and the full range of physiological adaptation. These major lacunae in our comprehension are a powerful argument for preserving species that may provide answers to some of the very significant remaining questions. As placental mammals, we have a special interest in our near relatives, and it is perhaps surprising to be so closely linked to such diverse creatures. A factor of 100 million is truly an awesome size range. When we can fully grasp its significance, we will perhaps better appreciate the full impact of "evolutionary potential."

The Smallest Free
Living Cell

There is a special fascination in thinking about the tiniest organisms that are capable of independent existence, the smallest living things. For many years, attention focused on the viruses as a logical limit; the advent of the new biology has made it clear that these particles are not capable of independent replication, but require the working apparatus of a host cell whose molecular machinery is used in virus assembly. The question then shifted to a consideration of the smallest free living entity, an organism capable of growing and replicating in the absence of other living forms. Seeking for such an organism has been an ongoing minor theme in present-day biological research. Of course, the quest has something of a never-ending quality about it, for having found the smallest organism yet described, one is never sure that an even tinier form may not be lurking in some unexplored recess. If we can ever develop a complete theory of molecular biology, then "the smallest free living cell" will become a theoretical construct to guide our search through nature.

An examination of contemporary microbiology shows that the smallest free living cells reported to date are among the order Mycoplasmatales. These mycoplasmas are a group previously called the pleuropneumonia-like organisms

(PPLO). While it is not possible to state the size with precision because the cells are quite variable in shape and dimension, the lower limit appears to be spherical cells about 0.35 μ in diameter. Let us try to get some sort of intuitive feeling for the significance of this very small size. The volume of such an object is 2.24 × 10^{-14} cm^3, or, alternatively, a liter of packed cells contains over forty million billion mycoplasmas. In terms of familiar microscopic objects, about 8,000 of these little organisms could fit inside a human red blood cell.

An alternative description of such minute structures can be obtained by comparing their size with the size of atoms. The approximate diameter of a carbon atom in biological molecules is about 1.4 × 10^{-8} cm, so that the smallest free living cells are about 2,500 atomic diameters across and are made up of about two and a half billion atoms organized into about six hundred million molecules, most of which are water. The solid part of the cell is made up of about five million molecules of amino acids, sugars, fatty acids, and other biochemicals. The units are organized into about ten thousand macromolecules, which are the working machinery of the cell.

We may get some insight into the size and complexity of a typical mycoplasma by conceptually magnifying it about ten million times until it just fits into an average-size office with a bit of crowding. Each atom is now a millimeter in size and each amino acid is about the size of a marble. Proteins are about as large as baseballs and the membrane is about as thick as the bathroom sponge. The structure is complex, but not so complex that we cannot envision understanding all the processes that are taking place and following them through the cell's reproductive cycle. The basic functions of life do not require an impossibly complex machine; they are within the scope of human understanding. The difficulty lies not so much in understanding the machinery, but in grasping the system rules that govern the interrelated behavior of the components.

These tiniest of living organisms can be analyzed from a

different perspective. Since all the genetic information of simple cells is encoded within a large circular macromolecule, DNA, we can examine this structure to determine the total of heredity information. Each of the cells we are considering contains a single molecule of DNA with a molecular weight of approximately four hundred and forty million daltons. This contains a coded message 730,000 characters long, written in a four-letter alphabet (adenine, thymine, guanine, and cytosine). If we wish to print the genetic message of mycoplasma in an ordinary-size book, it would require a volume about 360 pages in length. Thus, although the genetic instructions are complex, they are not so overwhelming that we cannot envision mastering them in a computer or some other man-made system.

The genetic information in the DNA is of course transcribed in an RNA sequence and then translated into a protein sequence. Since three nucleotides are required to encode one amino acid, the genetic message in polypeptides (a 20-letter alphabet) is 243,000 units long and could be printed in a small book of 120 pages. Because an average cellular protein is about 400 amino acids in length, the mycoplasma cell must have approximately 600 different kinds of proteins. Most proteins are enzymes and on the average we can assign a single function to each species of protein in the cell.

We are now at the heart of the matter. It appears that a system can carry out all the necessary chemistry, structure, and control to be a living cell at a level of complexity involving not over 600 individual steps. If we were to diagram the complex workings of this organism, we could set up a chart with 600 boxes and all the necessary interconnections. By characterizing the flows along these linkages we would come very close to a complete functional description. From a more numerical point of view, if we could relate the flows to the values of the quantities in the boxes we could formulate a computer representation that should map one-to-one on the activities of the cell. A program of 600 simultaneous equations is well within the range of modern computer technology. The smallest living cell is within the domain of a complete program. If the description is sufficiently complete, a laboratory experiment and an analogous computer print-out should yield the same results.

The preceding conclusions may come as a surprise to some. It had widely been assumed that thousands of steps were necessary to account for all of an organism's activities. Yet here is a cell that is biochemically almost as sophisticated as a mammalian organism and yet is able to carry out all its activities with 600 coded instructions. The path seems open to a rational and finite explanation of the minimal basis of life. Biochemical activities are relatively constant over the whole of the taxonomic kingdoms. If we can but understand the simplest cell we will well be on the way to comprehending what man is.

We are left with the residual question of whether or not there are cells smaller or simpler than the mycoplasmas. Our search through nature for microbes has not been an exhaustive one and our methods of enrichment culture will not grow every kind of cell. The question is experimentally an open one that can be pursued by a more extensive search. The general principles of molecular biology suggest that there is a lower limit to genome size somewhere around half that found in mycoplasma DNA. This estimate, which is a very rough one, is based on an examination of known biochemical functions that are assumed to be absolutely necessary for the operation of a living cell. In any case, in the study of mycoplasmas, we have come close to dealing with a minimal unit capable of persisting and replicating as a free living organism. At this level, the system is of finite and manageable complexity and indeed suggests that the substantial understanding of simple cells is well within our conceptual domain. Translating that optimism into actuality will undoubtedly take many years and much dedication.

Manufacturing a Living Organism

L urking behind many of the advances of modern molecular biology is the very real question of whether, within the foreseeable future, we shall be able to synthesize a living cell. Will it be possible to start out with water, graphite, sulfur, phosphorus, and other vital elements and emerge from the laboratory with an autonomously self-replicating biological organism?

There exists a class of experiments that relates to these questions in a way that has received little attention. These studies deal with the survival of living forms at temperatures near to absolute zero at −273° C. An extensive literature exists (see B. J. Luyet and P.M. Gehenio: "Life and Death at Low Temperatures," *Biodynamica*, Normandy, Mo., 1940) reporting on experiments in which living systems were taken to liquid hydrogen or helium temperatures, kept in that condition at a few degrees from absolute zero for varying lengths of time, brought back to room temperature, and placed under conditions of normal function. A number of forms including bacteria, seeds, and several kinds of microscopic animals survive this treatment with little or no loss of activity. In extreme cases, samples have been kept at a few degrees Kelvin for several days, following which the material was found to retain its viability. Rotifers, nematodes, tardigrades, and brine shrimp eggs are among the best-characterized examples.

To relate these low-temperature experiments to the synthesis of living systems, we must take the biological concepts of structure and function and examine them at the level of molecular physics. In describing an atomic particle for purposes of statistical mechanical analysis, we specify its state by three position coordinates (x, y, z) and three velocity or, more generally, momentum coordinates (P_x, P_y, P_z). Structure can generally be described in terms of the position-coordinates, and process or function corresponds to the momentum. Structure is the result of where the molecules are located, and function is a description of where they are going.

Temperature measures the amount of kinetic energy or the average value of the square of the atomic velocities as the atoms move about in a ceaseless thermal motion. When we reduce the temperature of a sample to its lowest possible value, all molecular motion ceases except for a small quantum mechanical zero point vibration, which need not concern us at this time. In reaching absolute zero, motion effectively comes to a complete stop and the system loses all memory of previous processes except those that have left a structural trace. In this state, the system can be completely specified by noting the positions of the atoms. In other words, all that exists at this temperature limit is structure; the concept of continuity of function loses meaning. Any organism that could survive near absolute zero would thus be immortal if maintained at that temperature, because one of the processes that would cease would be death or thermal degradation. This is the factor that some people have unsuccessfully attempted to exploit in the low-temperature storage of human corpses.

The fact that, for technical reasons, the experiments described above take the samples within a few degrees of the lower limit rather than absolute zero itself is not of significance in the present context, since the amount of molecular motion is negligible at the temperature studied. For example, at 2° K there is a vanishingly small probability of finding even one vibrationally excited molecule per cell.

When we heat the samples back to room temperature, we do not add back functional information, since heating is a disordering process that simply sets the atoms back in random thermal motion within the constraints imposed by the structure. Yet, the organisms, warmed to their growth temperatures, commence to function and carry out all their normal processes. The activity that arises is thus entirely the result of configurational information, contained in the system near absolute zero, interacting with its new environment.

The conclusion to be drawn from analyzing these low-temperature studies is that the synthesis of a living cell requires only that we construct an appropriate atomic configuration. If the structure is correct we do not have to worry about setting it running; it will set itself running. The problem of cellular synthesis is thus, in kind, no different from the problems of synthesis in ordinary organic chemistry. The difference lies in the fact that the structure is many orders of magnitude more complex than those usually made in the laboratory. The great simplification that emerges from the cryobiology experiments is the rather certain knowledge that the problem of synthesizing a living cell is within the domain of our existing conceptual framework. The task in principle merely requires a team of supersophisticated chemists adept at complex syntheses.

Consider then the hierarchy of chemical structures that go into making up a living cell. First there are the atoms, largely carbon, hydrogen, nitrogen, oxygen, phosphorus, and sulfur, which are assembled into small molecules, many of which are the monomers used in larger structures. The monomers are then combined into high-molecular-weight polymers, which associate by other than valence forces into organelles. The cell is an organized collection of these organelles with an intervening aqueous matrix.

Progress in synthesis has been made at all of these organizational levels. The construction of ordinary small-molec-ular-weight molecules from simple precursors is the task of organic chemistry, and it is generally believed that any stable, relatively small organic molecule can be synthesized if enough effort is expended. The joining together of monomers to form polymers has taken giant strides in recent years as ex-emplified by the synthesis of insulin, ribonuclease, and high-molecular-weight specific polynucleotide chains. The con-struction of organelles from macromolecules is a newer area of research. Yet, already it has been possible to separate small ribosome subunits into one type of ribonucleic acid and 21 types of protein that can be purified and subse-quently reassembled into active, functioning organelles. At the organelle and cell levels, microsurgery experiments on amebas have been most dramatic. Cell fractions from four different animals can be injected into the eviscerated ghost of a fifth ameba, and a living, functioning organism results. The ameba does not seem to require the detailed placement of each organelle; their presence appears to be sufficient. Thus, at every level of the hierarchy there is evidence that the structures can be assembled. The smallest living cells from the genus *Mycoplasma* are about 5,000 times as large as a single ribosome, but present results suggest that this is a difference of size rather than a difference of kind.

Of course, it would be premature to assume that we fully understand all of the structures, even in the smallest orga-nism. Many organelles are primarily built up out of mem-branes, and the nature of the protein lipid interaction in these structures is far from being completely understood. The molecular machinery involved in transport processes and various energy transductions has not been isolated, and our understanding of the detailed physical chemistry of these processes is the subject of an ongoing debate, even among the specialists. A great deal must yet be done to reduce our understanding of biological energy transformations to phys-ics and chemistry. Nevertheless, there seems little reason to doubt that these aspects of cell function lie within the theoretical domain of molecular physics.

There are those whose philosophical orientation toward the origin of life would have to be modified following the successful synthesis of a living cell. From the point of view of the present paradigm, the low-temperature experiments should already have forced a shift in viewpoint. Within existing molecular biology the synthesis of a living cell would not occasion any reorientation of theory, whereas the continuing failure to synthesize a living cell would force a reexamination of the physical foundations of biology.

Cell Types: The Great Divide

Some ideas are rapidly established in the consciousness of the scientific community, while others slowly infiltrate until they are fully accepted and the practitioners cannot remember their arrival date. The second type of historical development has characterized the notion that all living organisms are organized along either the prokaryotic or eukaryotic mode of cell structure. The fundamental dichotomy is at present an almost universally accepted empirical generalization underlying systematics, cell physiology, and evolutionary theory. Yet this distinction was not noted in the biology textbooks of the 1950s, scarcely mentioned in the 1960s, and is sometimes ignored in introductory works of the 1970s. While memories are still fresh it would be well for the historians of science to track down the growth and development of this most important concept.

Our concern here is not so much with the history as with some of the consequences of the idea itself. Prokaryotic cells are small, primarily unicellular organisms that lack membrane-bound internal organelles. The genetic material is usually contained in a single circular molecule of DNA. Eukaryotic cells are organized into membrane-bound compartments such as mitochondria, nuclei, chloroplasts, and lysozymes. The genetic material of eukaryotes is organized into chromosomes containing deoxyribonucleic acids and proteins. In general, the volume of a typical eukaryotic cell is one thousand times that of a typical prokaryotic cell. All of the conventional plants and animals are eukaryotes, as are the protozoa, fungi, and most of the phyla of algae. The prokaryotes include bacteria, blue-green algae, mycoplasma, rickettsia, and chlamydiceae (psittacosis group). The viruses are not included in this classification, since they are not cellular in nature. While prokaryotes are not very conspicuous, they are a major component of the ecologic cycling that takes place on the surface of the planet.

As the theory of prokaryotes and eukaryotes has matured, one salient and surprising experimental feature has emerged. All organisms are either one type or the other. There appear to be no cells that are intermediate in characterization between the two prototypes. This strict bifurcation provides a serious challenge to evolutionary theory. Four ideas have been proposed. The first view that emerged is that eukaryotes are descended from prokaryotes. A second view maintains that both are descended from a common ancestral type. Completeness demanded and got a third, though much smaller, school of thought, which maintains that prokaryotes are descended from the eukaryotes. A fourth and much more detailed theory, that of endosymbiosis, maintains that the organelles of eukaryotes are the evolutionary descendants of invading prokaryotes that have lost much of their independent function through a life of intercellular parasitism. These notions are at present subject to intensive discussion.

What embarrasses all of these theories and makes it difficult to choose among them is the absence of intermediate forms. Evolution is supposed to occur in slow stages with evidence of most or all intermediate forms. Thus our introductory biology courses tell us of the descent of the modern horse Equus from the Eohippus of 50 million years ago. However, we are shown the fossil evidence for the intermediate forms as well as vestigial evidence in the modern horse that indicate their development from the tiny earlier animals. No such obvious intermediates exist between the two cellular types. We are thus left with a number of unanswered questions. Why are there only two evolutionary stable modes of cellular organization? Why do we not find among the very many ecological niches any in which some intermediate form would have an advantage? What are the evolutionary relations between the two cell types?

The questions posed above are of more than purely academic interest. The last 30 years have witnessed a dramatic rise in our understanding of macromolecular synthesis. Most of the experiments have been carried out in prokaryotes and in particular in the bacterium *Escherichia coli*. Much of the present-day molecular biology is an attempt to take the knowledge derived from prokaryotes and apply it to problems of multicellular organisms. Thus, such problems as development and morphogenesis and cell and tissue control are being investigated by techniques and approaches that were so successful in *E. coli* experiments. Concepts such as genetic engineering and developmental control are also based on notions developed in the simpler taxon. Much of medical research is motivated by contemporary molecular biology. Thus, the question of the relation of the prokaryotes to the eukaryotes underlies a whole philosophy of approach to problems of pressing practical importance. How far will the prokaryotic paradigm take us in understanding mammalian systems? In biology, major theoretical issues are never very far from our ability to generate applications in medicine, agriculture, and related fields.

There are those who argue that the leap from *E. coli* to mammalian organisms is too large. They would suggest that we first turn our attention to simple eukaryotic forms such as yeast and fungi and use these organisms to bolster our basic conceptualization of the eukaryotic cell. They argue that such knowledge is a necessary prerequisite to developing a molecular understanding of very complex mammalian systems. Others feel that the biological generalizations are so deep that they supercede the distinction of pro- and eukaryotes. These researchers are directly involved in the molecular biology of the mammalian system.

Both groups agree as to the biochemical unity underlying biological diversity. The genetic code and the readout apparatus, including ribosomes, transfer RNA, and transfer enzymes, are common features of all cell types. The major biochemical pathways in energy processing and the production and utilization of ATP appear to be universal. The metabolic map and pathways to the synthesis of molecular building blocks also indicate a common ancestry for prokaryotes and eukaryotes. The differences begin to appear at higher levels of cellular organization, leading to multicelluarity and sexuality in the eukaryotes.

Whichever group proves to be more prescient, one prediction seems clear. The question of the relation of eukaryotic cells to prokaryotic cells will be with us for some time. When the answers are in they will illuminate the evolutionary history of life and provide some additional insights for the study of problems of applied interest. All life is divided into two parts and we know enough to be puzzled over the reasons why.

Pumping Iron

One of the surprises of modern biochemistry is the discovery of the incredibly sophisticated devices used by microorganisms to survive in inhospitable environments. It seems anthropomorphic to refer to these survival mechanisms as strategies, yet they so resemble some of the cleverest kinds of schemes we can devise that it is hard to find a more appropriate word. Indeed, when the study of life teaches new useful methods, we utilize the knowledge in bionics, the science of designing systems or instruments modelled after organisms.

An example of a cellular strategy is the mining of iron by bacteria. The element is one of the most plentiful on earth, composing about 5% of rocks. Because of the high atmospheric concentration of oxygen, most iron exists as very insoluble ferric oxides and hydroxides. Therefore, even given the substantial total quantity of the metal, the concentration in soluble form needed by organisms is extremely low, somewhere around one part per hundred million in the oceans. Nevertheless, iron is an essential element for most cellular processes, and a given bacterium can survive only if it somehow manages to extract such atoms from its environment.

The difficulties faced by microbes show some similarities to the technological tasks that have faced mankind since the beginning of the Iron Age. In both cases, the metal must be freed from a tenacious association with oxygen and converted into a usable form. The methods of resolving the problem are so radically unalike that they give insight into the basic differences between present-day engineering and biotic approaches. Commercial iron is produced by mixing ore (containing the oxides) with coke in a blast furnace where it is subject to a stream of very hot air. At the high temperature the oxygen combines with the coke, yielding reduced iron and carbon dioxide. The method is large scale, high temperature, and brute force.

Living systems have a much more subtle, low temperature, atom-by-atom method of solving the problem. The bacterium synthesizes and excretes into its surroundings molecules which have an extremely high affinity for binding iron at a very specific site. Such compounds are called chelators (from the Greek word for claw) because of the strength with which they scavenge every available free iron atom. This shifts the equilibrium and leads to the solubilization of the plentiful oxides. The next step in the process is carried out by a transport system in the cell membrane which pumps the complexed metal back into the cell. Within the cytoplasm the chelator retains the iron so that it is unavailable for biochemical reactions. Two different schemes are used to pry the iron loose from the molecular claw. In one of these, the enzymes digest the binding structure. This is a high price to pay, for the cell must synthesize replacement molecules to get more iron. In the other method, the iron is chemically reduced to the ferrous form which is then released. In either case, the metal atom is free within the cell where it is quickly utilized in making various ferroproteins. (A detailed account of these processes is found in *Microbial Iron Metabolism*, J. B. Nielands, Ed., Academic Press, 1974.)

As part of the policy of assuring adequate iron, some strains of cells may have two or more separate transport systems. An example is *Escherichia coli*. In the presence of citrate the cells have the ability to take up a ferric-citrate complex. Under low-iron conditions, they synthesize enterochelin, a chelator, and pump iron in the manner previously indicated. The genes for this system (induced enzyme synthesis) are expressed only under iron deficiency. These various functions involve a minimum of eight genes. Everything has its price, and the ability to survive low iron requires carrying a number of genes that are not otherwise utilized.

It appears that we have only begun to explore the uncannily clever devices invented by cells to circumvent iron depletion. Some lactobacilli utilize cobalt (coenzyme B_{12}), requiring ribotide reductase analogous to the iron-containing enzyme of most other bacteria. *Clostridium pasteurianum*, when grown on very low iron, synthesizes a noniron-containing flavodoxin with properties similar to ferrodoxin. There are suggestions that molybdenum replaces iron in some metalloproteins.

One of the most fascinating suggestions from the study of metal transport is that *Salmonella typhimurium* is capable of "scavenging" iron complexed with molecules secreted by other organisms. Here we have biological competition reduced to its most atomic level. We tend to envision survival of the fittest as a matter of tooth and claw; not necessarily so. Imagine a number of bacteria growing in a soil where the limiting factor is the amount of available iron. A struggle ensues to see which gets those few atoms. Each species begins to excrete its own specific chelators, and the advantage goes to those organisms whose molecules win out in binding the iron. The race is neither to the swift nor to the strong but to the highest binding constant. Thermodynamics and free energy of interaction have replaced tooth and claw in the battle for survival. Now along comes a microorganism that is losing the race for iron binding. A series of mutations lead to a transport system that can pump the high ferric-binding molecules of another species faster than can the original species. It is clearly theft, like stealing another man's lobster pot, but evolution is determinedly amoral, and the second species, which lost the thermodynamic battle, wins the kinetic war by pumping iron faster than the competition. Thus it is in nature at every level and often by devices so subtle that we can just begin to glimpse them.

It is a bit unnerving to pick up a handful of soil and to think about the kinds of battles that are being waged and the molecular ingenuity that is often involved. It requires some orientation of outlook to shift thought from that Darwinian view of antelope being chased by lions to the calmer, but no less deadly, struggle taking place at all times. Of course we are not totally removed from this particular situation. Iron-binding proteins in the bloodstream keep the concentration low enough to make healthy circulating body fluid a poor medium for bacterial growth. We, no less than our microbial associates, are engaged in the business of pumping iron and holding it in the right molecular configuration. So it goes for the relatedness of all life.

On First Looking into Bergey's Manual

I f microbiologists were given to poetry, they could sit back and expect someone to pen "On First Looking into Bergey's Manual" to match John Keats' "On First Looking into Chapman's Homer." The occasion for this romantic thought was the arrival in my laboratory of the eighth edition of *Bergey's Manual of Determinative Bacteriology*, edited by R. E. Buchanan and N. E. Gibbons (the Williams & Wilkins Company, Baltimore, 1974). The volume appeared while I was away and the package was opened by my research associate, who all but shouted "Great, the new *Bergey's* is here." An undergraduate research student, being young enough to have missed the impact of the seventh edition, wanted to know what was so great. I returned shortly after and seeing the book chortled in glee that "The new *Bergey's* is here." By this time the undergraduate was beginning to get the message.

The first edition of this indispensable guide to the systematics of bacteria was published in 1923, and there has not been a revision since the seventh edition appeared in 1957. During the intervening 17 years up until the present, molecular biology has become an important science in its own right and bacteria have played an important role in the development of that science. Curiously enough, the vast majority of studies have been carried out on only one species,

Escherichia coli, and bacterial taxonomy has not been a subject in the limelight. Nevertheless, important findings have been made over the entire range of bacterial species, and the new edition of *Bergey's Manual* can be expected to serve as a crucial tool in our broadening understanding of prokaryotic cells and the interaction of certain species with human hosts.

In classifying bacteria, the eighth edition divides these microbes into 19 broad phenotypic groups and subdivides the groups into genera and species. In some of the major divisions the genera are grouped into families, but no attempts at grouping into higher taxa are made. This represents a recognition that the post-Linnaean system, which has been of such value in dealing with the sexual eukaryotes, does not correspond to the factors responsible for speciation in prokaryotes. Indeed, bacterial taxonomists are constantly plagued with the problem of what constitutes a species. This is perhaps reflected in the present distribution of species among genera, ranging from one species per genus for 100 of the bacterial genera to the genus streptomyces with 463 listed species. This may say as much about the classifiers as the organisms or may, in fact, be a tribute to the economic importance of the range of antibiotics that are produced by various streptomyces. In any case, the manual lists 245 genera, 1,576 recognized species, and a number of other *species incertae sedis*.

An introductory perusal of the new *Bergey's Manual* reawakens an old question. Why are there so few species of bacteria? In the essay "Homage to Santa Rosalia or Why Are There So Many Kinds of Animals?" the ecologist G. E. Hutchinson postulates that there are so many species because there exist so many niches, microenvironments in which different phenotypes can have selective advantages. Thus as the size of animals decreases, the number of available microenvironments increases and the number of coexisting species increases. For bacteria, the number of microniches is huge. For example, a coring of soil shows, microscopically, a very wide variety of chemical and physical properties. So we are forced to consider the question of why there are so few species of bacteria. Indeed, why are there species of bacteria at all? Given that each bacterial species has several thousand genes in its hereditary apparatus, each capable of independently mutating and giving rise to many alleles, the number of available genotypes is almost limitless. Since organisms that reproduce by binary fission have no gene matches required for mating, any biochemically valid set of genes should give rise to a viable bacterium. We might indeed expect a continuum of types rather than well-recognized species. What is revealed by the studies that have gone into *Bergey's Manual* is that there are, without question, well-recognized species, small in number and rather ubiquitous in occurrence. Certain constellations of genes appear to produce stable prokaryotes that can outdo the competition. The question of speciation in bacteria, nevertheless, remains something of an enigma.

In part, of course, the reason for the small number of species may lie with the experimenters. We may have been insufficiently imaginative to have devised culture methods appropriate to the whole range of microorganisms. There may yet be lurking in nature whole major groupings waiting to be discovered. Alternatively, there may be lurking in the information in *Bergey's Manual* principles of prokaryotic organization that could lead us to a much deeper understanding of the living cell.

From a less abstract point of view, the eighth edition gives us a new summary of the properties of several species that live in association with humans. We have already mentioned *Escherichia coli*, which is ubiquitously found "in the lower part of the intestine of warm-blooded animals." This gram-negative facultatively anaerobic rod has gained a permanent niche in biological science. Accompanying almost all humans is *Propionibacterium acnes*, which is "isolated from the normal skin (approximately 10^4/sq in)" and is the most common contaminant of anaerobic cultures. Another coinhabitant of our bodies is *Veillonella parvula*, which is "parasitic in the mouth and in the intestinal and respiratory tracts of man and other animals." These three species must occur at a level of 10^{12} cells or greater in and on the average human body.

The disease-causing bacteria are all catalogued, not by illness but by their bacteriological properties. In addition to the usual bacteria, part 18 deals with the rickettsias and part 19 covers the mycoplasmas. The rickettsias are morphologically like bacteria but are obligate intracellular parasites. The mycoplasmas lack cell walls and are still something of a taxonomic anomaly.

Bergey's Manual is, of course, a fascinating repository of facts. One is able to learn, for example, that *Rickettsia tsutsugamushi* derives its species name "from two Japanese ideographs transliterated *tsutsuga*, something small and dangerous, and *mushi*, a creature now known to be a mite." One is also able to learn that acne is an incorrect transliteration of the Greek noun *acme*, meaning a point. There are opportunities to become familiar with *Thiobacillus thiooxidans*, which causes sulfur broth to become more acid than pH 1, and *Sulfolobus acidocaldarius*, which has a maximum growth temperature of 85°C! One can muse over the observation that *Lactobacillus bulgaricus, Lactobacillus jugurti*, and *Lactobacillus lactis* can occur simultaneously in yogurt, or, surprisingly, that *Treponema pallidum* and *Treponema pertenue*, the etiological agents of syphilis and yaws, have still not been cultured in vitro.

As fascinating as I find the eighth edition of *Bergey's Manual*, I do not expect to find it on best-seller lists. It is rather heavy to hold for reading in bed at night. While it is rich in cast of characters, the plot is often confusing and the underlying themes are mixed. But if one wants to read bits and pieces and wonder about the eternal query "What Is Life?" then this work is a bountiful storehouse of the results of hundreds of thousands of man-years of work. It is truly an achievement worthy of a poetic accolade.

Stalking the Heffalump

While *Winnie-the-Pooh* by A. A. Milne hardly qualifies as a biological monograph, I suspect that a species count would reveal at least 20 entries, both real and imaginary. Granting that a score of taxa is considerably fewer than the 3,800 that ecologist C. Elton reports in the same kind of English countryside as Christopher Robin's play area, it is 19 more varieties than many biologists get to work with in a lifetime of research. One of the great potential sources of edification is the existence in the biosphere of more than 1,700,000 species, many of them ideally suited to answer some fascinating scientific questions if we but knew how to address the queries to the organisms. It is one of those all too human limitations that a given individual can get to know but a small fraction of this multitude of plants, animals, and microbial forms. Aristotle's writings on biology refer to about 500 diverse species, but the time has long since passed when a scientist could cover such a range of material. Some devote an entire career to a defined aspect of a single variety of organism.

As one usually occupied at a desk or laboratory bench, I have welcomed opportunities to get out into the field seeking new biota of interest. Although the creature may be common or rare, locating it in its own habitat and getting to know it better is a rewarding experience. One of my first chances at this kind of endeavor came many years ago when our attention turned to the tardigrades or, as they are more commonly called, "water bears." The concern with these tiny, almost microscopic, eight-legged animals related to their ability to withstand high vacuum and temperatures near to absolute zero with no apparent ill effects. The literature reported that specimens allowed to dry at room temperature and kept in that state for 18 months returned to life about two hours after having been transferred into water. This drying seems most remarkable when we note that these animals have a highly developed nervous system with a cerebral ganglion and a number of other nerve centers. The osmotic and ionic imbalance that these neurons suffer on desiccation must be overwhelming. The ability of the organism to survive this kind of trauma, indeed to flourish under it in nature, serves to challenge our understanding of a number of well-established principles of physiology. Here was a beast worthy of the scientists' attention.

Armed with a low-power microscope and accompanied by my children, I went seeking the tardigrades in the woods in back of our house. Only when we returned empty-handed after several unsuccessful expeditions did my offspring accuse me of "stalking heffalumps." However, biology is an art as well as a science and a naturalist must know where and how to look for his animals. Thus I was pleased some months later to be in North Carolina in the presence of one of the country's few authorities on these unusual fauna. We were walking through an estate and came upon a gargoyle fountain. Where the spray hit the surrounding rocks moss was growing and my mentor announced: "That growth will have tardigrades." Samples were taken, and sure enough the microscope revealed many water bears feeding upon the moss.

Emboldened by my instruction from the expert, I returned home, gathered my family, and marched back into the woods to the waterfall. Where the spray hit the rocks, moss was growing, and I proclaimed: "There are tardigrades on that moss." We hurried back with scrapings from the wet stones and placed them under the microscope objective. To my great relief a number of the sought-for animals came into view and I avoided being thought of as "a Bear of No Brain at all." Although our own tardigrade research never developed, we did interest others in this study, and my feeling for the richness and variability of animal forms was immeasurably increased. It is really quite an experience to desiccate an animal for many days and then place it in a drop of water and watch under the microscope as it swells, begins to move, and walks off.

An entirely different type of exploration was occasioned by focusing attention on the question of what is the smallest autonomous self-replicating organism. We sought a unicellular form that was capable of growth in a broth medium free of other cells that might serve as hosts. Here we did not go into the field, but took to the postal system and wrote to microbiologists the world over asking for their tiniest free-living microbes. The pleuropneumonia-like organisms, which were then known by the acronym PPLO and have since become mycoplasma, emerged from our investigations as the ideal objects of study from the point of view of small size. At that time the PPLO were almost exclusively studied by veterinary pathologists. These tiny cells proved useful in a number of basic biological investigations, all of which were confined to the laboratory. Mycoplasma helped to establish the minimum hardware necessary to function as a cell. For here is a living organism whose entire body consists of only about one billion atoms. We are truly observing the domain of overlap between biology and atomic physics.

A more recent probe for a proper organism found me snorkeling in the channels of San Diego harbor looking for octopus infected with dicyemid mesozoa. The mesozoans are a truly enigmatic group, so different from other forms that many zoologists regard them as a quite separate branch of the Animalia. They are tiny worm-like creatures containing only about 30 cells in the adult stage, which consists of a central reproductive cell surrounded by a coat of somatic cells. The animal world is comprised mainly of the single-celled protozoans and higher animals which contain at the very least several thousand cells. The only exceptions to this rule of "one or very many" are the mesozoans; they alone form stable organisms with an intermediate number of cells. As an exception they are a puzzle, although the rule itself is a little-discussed enigma of modern biology.

What seemed strange on first looking into this area of zoology is the existence of this separate grouping within the subkingdom of multicellular animals that is virtually ignored by modern researchers. In the midst of a great growth in the literature of the life sciences, only a few papers a year deal with this taxon. Yet because of their very difference from all other forms they must have much to tell us of evolution, systematics, and developmental biology. The issue was settled; we set out to study the potential of mesozoans as laboratory organisms. The first practical problem in studying the dicyemids is that they are found almost exclusively as kidney parasites of the cephalopods (octopus, squid, and cuttlefish). As I dove down to try and grab a fast-moving octopus the thought did occur to me that there must be easier ways of asking fundamental biological questions. However, these fleeting doubts were later resolved upon looking throuvh the microscope and observing a creature so different in basic tissue organization and body plan.

The mesozoan story is still unfinished and likely to remain so for several years, but a side benefit of the study is an appreciation of the octopus. Seeing them on the ocean bottom or caring for them in tanks gives some insight into a highly developed invertebrate system. It is difficult to penetrate the thoughts of an animal unlike ourselves in structure, physiology, and habitat. Octopi have a complex behavioral repertoire and clearly are fascinating animals in their own right. A case in point, which we rapidly rediscovered, is the considerable difficulty in keeping specimens from escaping their tanks.

The 100 best-studied organisms must no doubt account for well over 90% of all the physiological and biochemical research that has been carried out. That leaves a lot of creatures out there that are known by name alone, and a lot of others that are as yet undiscovered. As I have pointed out before, both analytically and descriptively we are just at the beginning of our knowledge of the biological world, which should be heartening news to young researchers. Why, in addition to heffalumps and tardigrades and mycoplasma and mesozoa, there are piglets and rabbits and owls and woozles (or wizzles, as the case may be!). If we can keep Winnie-the-Pooh's sense of wonder, we have a lot of surprises yet to come.

PART IV

Zen and the Art of
Getting into Medical
School—COMMENTS ON
EDUCATION, RESEARCH, AND
MEDICINE

Zen and the Art of Getting into Medical School

In the provocative book *Zen and the Art of Motorcycle Maintenance*, Robert Pirsig tells the story of the gifted scholar Phaedrus who went insane trying to define "Quality." Each year 119 medical school admissions committees are similarly required to define "Quality." The issue is not an abstract one, but an immediate, pressing, tangible, existential reality, for each committee must select a small number of matriculants from a large field of applicants. Putting aside for the moment the difficult logistic problem of multiple applications, there are approximately 45,000 candidates for 15,000 positions. Taking into account the substantial barriers that students face in becoming candidates, I would guess that there are about 40,000 qualified applicants, among whom over 30,000 are well qualified. Thus the admissions committees are not free to defer the question of "Quality." They must arrive at an immediate, pragmatic resolution of the issue.

In quite another way, *Zen and the Art of Motorcycle Maintenance* reflects on the problems of admissions committees. For the dichotomy in the book's title also refers to the tension between the rational, scientific approach and the arational, artistic, holistic approach to life. It is an expression of a dominant modern theme of the two sides of the brain: the intellectual, verbal, causal left hemisphere and the sensuous, spatial, intuitive right hemisphere (see, for example: R. E. Ornston, *The Psychology of Consciousness*, Penguin Books, 1972). And so committees survey the candidates. There is the grade point average, Medical College Admission Test, organic chemistry mark—left-hemisphere assessment. For candidates who clear this barrier, there is the interview, extracurricular activity, letters of recommendation—right-hemisphere assessment. The problem of how to weigh the two hemispheres is one that is capable of creating great conflict among admissions officials. "Do we want the history major who plays flute in a woodwind quartet or the biochemistry major who has published two papers on nuclear magnetic resonance spectroscopy, or the psychology major who has maintained a 3.2 average while playing varsity hockey and doing volunteer work in the pediatrics ward?"

One suspects in the end that among the approximately 1,800 members of admissions committees there is a very wide range of definitions of "Quality." If all of this is likely to drive the committee members Phaedrus-like into electroshock therapy, the responses of the 45,000 candidates for admission are even more complex. The students know they are going to be judged on "Quality," but for some there is the Kafkaesque experience of having no one to tell them what "Quality" really is. The president of the university informs them that marks are not the sole judge of a man or woman. At the same time the premedical adviser warns them that unless they have a certain minimum average, no medical school is going to look at them. The coach points out that character on the field counts, while the chemistry professor stresses that undergraduate research is what makes schools take notice. The dean tells them of the importance of community service activity, while the laboratory instructor stresses that identifying the unknown is the be-all and end-all of human perfection.

Some students respond to these schizophrenic messages in fairly predictable ways. The most success oriented try to deduce how to maximize their "Quality" points among the various left- and right-hemisphere activities. They are the "wunderkinder," and an individual's credentials may show varsity fencing, second violinist in the student symphony, summer research on hormone action in worms, barbershop quartet, debating society, volunteer affirmative action, chess club, summer employment in an advertising agency, and a bewildering variety of other activities. Another group of students devote themselves to the left hemisphere with a vengeance, and emerge with 3.99 grade point averages and no interpersonal interactions with anyone in the university. And so it goes among the possible strategies.

In the midst of all this confusion, there is the cry that the stress on premedical training is ruining the character of liberal arts education because of the narrowness of the endeavor and the competition leading to a variety of excesses. The issue is not a simple one. If we will agree that the essence of a liberal education is the unfolding of one's own sense of "Quality," informed by tutors, exemplars, and the writings of others, then we must assume that there will be many pathways to a liberal education. Breadth or narrowness is not the sole criterion; rather it is the search for meaning in a student's development. The danger of premedical training is that early in life an individual will surrender his personal, internal search for "Quality" and adopt a mask designed to impress admissions committees. The individual becomes all facade and no substance, a con artist at getting marks and letters of recommendation, but beneath it all a hollow human being. I feel that our medical schools have more than their fair share of such people. They are hard driving and success oriented, but one feels they may lack the ethical or intellectual dimensions of excellence. They have been unwilling even to listen for the beats of their own drummers. Those who have not worked out their own sense of self often show the greatest distress at the thought of not being admitted. As a reaction to such people, I have often been tempted to put a sign on my office door reading: "Anyone who is panicked about getting into medical school is unfit to be a physician."

In the end, the failure of admissions committees to adopt a uniform standard of "Quality" is probably the ultimate protection of a liberal education. For the existence of a precise image would provide an exact model for those willing to assume the mask of "Quality" and would disadvantage those who were seeking for their own sense of excellence. The situation of not being presented with a narrow set of right-hemisphere requirements will be severely traumatic only for those unwilling to undergo their own search for standards, that is for those unwilling to participate fully in a liberal education. What is being discussed here for the premedical situation is also true of the educational system in general. Early evaluation depends almost entirely on the well-defined left-hemisphere assessment. Only under the protection of a true liberal arts college can an individual have the leisure to mature in a balanced sense.

In part, today's problems are with the admissions procedure. The overwhelming number of applications makes it very difficult to examine in full depth the dimensions of the candidates. The formal quantitative assessment is relatively simple and is probably given greater weight. The right-hemisphere assessment forces members of admissions committees to come to their own conscious definitions of "Quality." And as we learned from Phaedrus, defining this concept can be a very dangerous procedure indeed.

Competition will always be with us. Without endorsing social Darwinism, one must acknowledge that competition is part of our biological heritage. Rather than eliminating it, we must direct attention at what is being selected for. The evolution of medicine is in the hands of admissions committees. It is a heavy responsibility. We need procedures for the competition to make it more responsive to values and, hopefully, better. There is a modest suggestion that I believe could be a step in the right direction: If students were encouraged to limit their applications to perhaps six, then the total number of pieces of paper to be processed by admissions committees would drop drastically, and the members could devote more effort to individual decisions. In addition, the students would focus on where they really wanted to go and realistically feel they have a chance of acceptance. The present buckshot procedure of trying to up the odds by throwing the dice the maximum number of times hardly encourages thoughtful selectivity. We are searching for very human attributes; we must somehow reduce the task to human proportions before much progress can be hoped for. It goes without saying that a few thousand more seats at medical schools would also help.

The role of the physician in society is undergoing great change. Forces clearly are operative that will drastically alter the practice of medicine over the next few decades. One of the great determining factors lies in the hands of admissions committees, for they determine the character of the individuals who will participate in the decision making.

The Merck of Time

Several years ago, our oldest son was leaving for a term in the Peace Corps in a remote and underdeveloped country. In discussing which items to crowd into an overstuffed backpack, we suggested *The Merck Manual of Diagnosis and Therapy*. That single volume went on to become a reference library, a warming bit of Americana to an entire cohort of volunteers on a group of South Sea islands. The Manual belongs to that genre of writing that manages to condense an entire complex field into a single book, albeit one of 1,964 pages (12th edition). Incomplete as such a condensation must inevitably be, it represents a declaration of the state of the art at the time of publication. Therefore, it was specially interesting to lay hands recently on a first edition of this reference work.

Imprinted in large type on the title page were the words *Merck's 1899 Manual of the Materia Medica*, followed by the statement "Together with a Summary of Therapeutic Indications and a Classification of Medicaments." The price was one dollar and the motto *Multum in Parvo* was an apt description of a fact-packed book that readily slips into my side coat pocket with barely a bulge. A glance at this work led to a perusal of the other 11 volumes and the realization that within these editions is written the pragmatic history of medical practice from the beginning of its modern epic at the turn of the century to the highly sophisticated stage we believe we see at the present. However, tracing out that path is a task we will leave to medical historians, contenting ourselves now with a brief look at the first edition.

The Foreword describes the book's function in a delightfully humble manner: "Memory is treacherous. . . . In Merck's Manual the physician will find a complete Ready Reference Book. . . . A glance over it just before or just after seeing a patient will refresh his memory in a way that will facilitate his coming to a decision. . . . The publishers may be allowed to state that they have labored long and earnestly, so to shape this little volume that it shall prove a firm and faithful help to the practitioner in his daily round of duty."

The subject material is arranged in three compact sections: the Materia Medica, Therapeutic Indications, and Classification of Medicaments According to Their Physiologic Actions. The Materia Medica surprisingly lists over 900 items, from "Absinthin (in anorexia, constipation, chlorosis, etc.)" to "Zinc Valerianate—U.S.P. (diabetes insipidus, nervous affections, neuralgia, etc.)." The list is heavy in inorganics and natural products and sparse in synthetic organics, which are so much a part of contemporary medicines. However, it must be remembered that only 70 years had intervened between the very first organic synthesis and the compiling of this list. As might be expected, many names in the list have totally disappeared from the modern pharmacopoeia. For example, one wonders about an item "Aletris Cordial (Prepared from Aletris farinosa or True Unicorn)."

Part II, Therapeutic Indications, first provides a list of diseases. Then we find the materia medica and other agents used in the treatment of these conditions. "The statements hereon are drawn from standard works of the leading modern writers on Therapeutics and supplemented—in the case of definite chemicals of more recent introduction—by the reports of reputable clinical investigators." All of the items in this part therefore had some standing in the field.

One finds some drugs surprisingly widely used. An example is arsenic, which is listed under 105 diseases, including arthritis, bronchitis, cholera asiatica, diabetes mellitus, elephantiasis, fever, gastritis, hydrophobia, impotence, jaundice, keratitis, and so on down the alphabet. Other widely used drugs included belladonna, cannabis indica, eucalyptol, iodine, and mineral acids.

As late as the turn of the century, bloodletting had not disappeared as a therapeutic device and was suggested for several conditions. In acute bronchitis one of the listed procedures was "Bleeding, from the superficial jugular vein in severe pulmonary engorgement." Bleeding was also an indication for headache, insomnia, intermittent fever, pericarditis, pleurisy, puerperal convulsions, pyemia, spinal concussion, and sunstroke. The medicine of 1899 was in some stage between medieval concepts and modern theories.

It seems strange that the renaissance occurred so late in a field of such importance to individuals, but we may recall that bacteriology and organic chemistry did not develop until the second half of the 19th century, and biochemistry had barely started at that time. The basic sciences had not yet formulated a rationale for therapy, and pragmatic approaches were used, although the correlation between treatment and result must have often been in the minds of the physician and patient. There is, after all, a natural feeling that it is better to do something than to do nothing. For example, in the case of diphtheria and gonorrhea, well-defined bacterial diseases for which no cure existed in 1899, there were 75 and 96 possible treatments listed. Some of these must have provided symptomatic relief and others were quite toxic, yet they gave the physician a large available armamentarium. In contrast to the first edition, the 12th notes essentially one course of treatment for each of these diseases.

The metabolic disorder diabetes mellitus was well recognized and 68 specific treatments were given. One begins to see a pattern. The less a disease was understood, the larger the number of treatments available.

The third part of the manual, Classification of Medicaments According to Their Physiologic Actions, is an eight-page list that is quite short compared to the wide variety of therapeutic indications. It does represent a firm commitment toward a rational approach to therapy based on specific physiological effects. We can even see the emerging relation of medicine to biochemistry under the heading "Digestives," where diastase of malt, pancreatin, papain, pepsin, and peptenzyme appear.

Considering our usual view of the late 1800s in terms of the Victorian ethic, a surprising fraction of the book is devoted to explicit sexual disorders including chordee, emissions and erections, exhaustion-sexual, impotence, nymphomania, satyriasis, and spermatorrhea. The classification section lists 10 aphrodisiacs and 16 anaphrodisiacs. Among the cures for nymphomania is "Tobacco: so as to cause nausea: effectual but depressing."

It is easy to sit back and chuckle at our predecessors. Hindsight is a most effective intellectual tool. But I think that amusement alone would be a poor use to make of the 1899 Manual. The practitioners of that era were no less dedicated and considered themselves no less educated than those of today. In 1871, a full 28 years before this first edition, Oliver Wendell Holmes was able to say to a medical college graduating class: "You are now fresh from the lecture room and the laboratory. You can pass an examination in anatomy, physiology, chemistry, materia medica. . . ." The physicians of 1899 regarded themselves as neither barbers nor butchers, nor did they regard themselves as ineffectual. They were highly trained professionals, medical scientists, applying the methods of their learned professors to problems of human suffering.

Perhaps, then, we might project ourselves 77 years into the future and envision some curiously perverted prose writer thumbing through a yellowed, cracking 12th edition of *The Merck Manual*. Surely he too will have a few laughs and wonder how ineffectual was the physician of 1976. He will list diseases for which there were many treatments and no cures. He will find much of the language quaint and outdated, and he will be surprised at how little we actually knew of the physiological and biochemical mechanisms of health and disease. The message of this small first edition is thus not pride, but humility. There is so much to know of the human condition and we understand so little. *Allons!* Back to the laboratory bench, the clinic, the library. Let that prose writer of 2053 know that at least we tried.

Helmholtz, Mayer, and the M.D.-Ph.D. Program

The last few years have witnessed a proliferation in the number of joint M.D.-Ph.D. programs offered in American universities. The purpose of these double doctorates is presumably to train a select group of students in preparation for a career in medical research, combining an understanding of modern medical practice with an in-depth ability to use the tools of modern physiology, biochemistry, cell biology, and biomedical engineering. The 19th century and early 20th century offer numerous examples of individuals trained primarily in medicine who made significant contributions to the basic sciences. Indeed, the most profound principle of physics, the conservation of energy, was very largely formulated by two physicians, Julius Robert Mayer and Hermann von Helmholtz. The careers of these two men give us some insights into the problems of combining research and clinical activities.

Mayer, who was born in 1814, studied medicine at the University of Tübingen and received his doctorate of medicine in 1838. In 1840 he signed on for a year as physician on a Dutch merchant ship. While in Java, certain physiological observations associated with blood letting convinced him that there was a single unifying indestructible force in nature. He recognized the outlines of the relation between motion and heat and wrote a paper that was rejected by the editor of *Annalen der Physik und Chemie*. Indeed, Poggendorff, the editor of that esteemed journal, had the distinction of rejecting both Mayer's and Helmholtz's first fundamental papers on the conservation of energy. Editors and referees take heed of the judgment of history.

Mayer settled in Heilbronn, married, established a successful medical practice, and devoted considerable effort to extending and elaborating his scientific ideas. In 1842 his more definitive paper was published in Liebig's *Annalen der Chemie*. This paper established his claim to priority in formulating the conservation law. In the years between the mid-1840's to 1860, Mayer formulated physiological theories on animal heat, extended his conservation hypothesis, and carried out significant studies in astrophysics. His life was upset by a depression in part over failure to achieve recognition and in 1850 he attempted suicide. He spent some time in mental institutions. In 1858 recognition came to him largely through the influence and prestige of Helmholtz, who argued Mayer's priority in establishing the conservation of energy.

Mayer remained, however, an outsider to science. He was an idea man who did no experiments. He was driven by philosophical notions to formulate a conservation theory before he fully understood the physical and mathematical basis of what he was doing. It is this particular feature of his creativity that makes it so difficult for us to comprehend the psychological factors. In his early career before his trip to Java he apparently showed little hint of his creative genius. He was a mediocre secondary school student and was once expelled from the university. It is doubtful that an individual of his character and background would be admitted to a current M.D.-Ph.D. program.

Hermann von Helmholtz, on the other hand, achieved preeminent success within the scientific community. He is regarded by some as the last universal genius, the last Renaissance man. Born in 1821, Helmholtz studied at the Potsdam Gymnasium and wished to study physics at the university. His family could not afford to send him, so he accepted a five-year state grant to study medicine in Berlin. In return he agreed to an eight-year service as an army surgeon. This early fellowship arrangement allowed him to continue his education. At Berlin he was able to pursue studies in chemistry and physics at the university, while he completed his M.D. degree in 1842.

Fulfilling his earlier agreement he became assistant surgeon to the Royal Hussars at Potsdam. His official duties were light and he built a small laboratory for physics and physiology in the barracks where he worked on muscle heat and the rate of transmission of nervous impulses. Helmholtz was associated during this period with duBois-Reymond, Brucke, Ludwig, and Muller, men who laid the foundations of modern physiology. Between 1845 and 1847 the young surgeon worked on the conservation of energy and carried out some of his most productive work while on field hospital duty. Unaware of the work of Mayer, he formulated a much deeper, more sophisticated version of the conservation law. Although he had been stimulated to think of energy by his physiological studies, his 1847 paper revealed him a master of theoretical physics. Our modern formulation of this most basic principle of physics follows directly from that presentation.

In 1848 he was released from military service to assume a chair in physiology at Königsberg. His subsequent academic career is a review of mid-19th century science. He invented the ophthalmoscope in 1851 and in 1856 published the first volume of his monumental *Handbuch der Physiologischen Optik*. He published notable works on acoustics, music, hydrodynamics, electrodynamics, thermodynamics, and epistemology. In 1871 he assumed the last of his academic positions, professor of physics at Berlin. It is difficult to generalize from the career of a genius like Helmholtz, yet he is clearly the type of individual who would be eagerly sought in the current M.D.-Ph.D. programs. He worked fully within the scientific establishment of his time and reached beyond the frontiers in numerous areas.

In examining the current programs, one aspect of these two striking careers is worth noting. Helmholtz and Mayer were each 26 years old when they wrote their respective major contributions on energy conservation. They were each several years out of medical school and each had a position that provided the leisure to pursue the deep intellectual pathways that had been chosen. Today's M.D.-Ph.D. programs keep individuals totally comitted until age 30. There is no leisure to pursue the scientific byways or to attempt unorthodox approaches to problems. What are normally the most creative years are spent in programs that are totally occupied with requirements of one sort or another.

Now this total education for the first 30 years of life may produce individuals superbly trained to function within the paradigm, but does it also inhibit the creativity of those few catalytic individuals who are able to carry science and medicine into new domains of thought? I am raising the question of how much training is too much within the context of the human lifespan. I am also raising the question of the inhibition of human creativity by insufficient leisure. No answers to these questions are being proposed here. Rather the assertion is made that they are valid concerns for the constructors of programs that are getting progressively longer. An occasional look back to our intellectual predecessors may provide some guides. Mayer and Helmholtz must certainly be ranked among the most successful medical school graduates of all time.

Sacred Cows and Sacrificial Guinea Pigs

I t is strange that when we are reading, a single word or phrase often catches our attention and sets the mind off on a distracting journey into free association. Thus my attempts to become informed on physiology and animal biochemistry are frequently interrupted when the phrase "the animals were sacrificed" jumps out of the paper and impinges on my retinas, which then send signals whirling around the cerebrum, switching whole series of neural circuits. After enduring this phenomenon for many years, I decided to track down the cause of this idiosyncracy and root out, at the id level if necessary, this deterrent to my education.

The cause of the problem, of course, went far into my childhood and stemmed from my mother's urging that when I did not know a word I go to the dictionary and look it up. Somewhere in that past I had found out a *sacrifice* was "primarily, the slaughter of an animal as an offering to God or a deity." This seemed at first sight to be very different from the procedures so casually described in articles on animal experimentation in biochemistry, physiology, and pharmacology journals. An effort to elaborate the difference took me to the encyclopedic notion that "Sacrifice is a religious rite in which an object is offered to a divinity in order to establish, maintain, or restore a right relationship of man to the sacred order" (*Encyclopaedia Britannica*). The topic of human and animal sacrifices has been the subject matter of anthropologists, social psychologists, and religious scholars. I perused this literature in an effort to gain some insight into the modern use of the word and was led id over ego by free association to the following whimsical view of a contemporary anthropologist writing on the subject of the sacrifice of experimental animals.

"Animal sacrifice had all but died out in the established religions of the western world, when in the 1800s it mysteriously reappeared among a new cult known as *Physiologists*. In order to comprehend the activities of this sect we must note the similarities and differences between the new elite and their historical predecessors. The Physiologists practice their rites in cultic centers called *Laboratories*. These structures differ from the older Temples in that the sacrificial altars which are the center of the public rooms in the Temples are found in small private chambers in the Laboratories. There is often a single large public room in the Laboratories, but it is restricted in use to one-hour devotional meetings where the congregation sits in the dark while a chief sacrificer describes his rituals. These sessions are called *Seminars*, and sometimes the devotees become so absorbed in the ritual that they are seen to close their eyes in prolonged meditation.

"Membership in the cult of Physiologists is very select, and an eight- to twelve-year period of apprenticeship is required. The training consists of memorizing the sacred paradigms and learning to carry out sacrifices with scalpels, needles, electrodes, and various strange potions. The entrails and blood of the victims are then placed in very varied ways into large machines. The apparent purpose of processing the blood and organs in these elaborate modes is to read omens of the future. The novitiates must learn to operate the portent machines, and they are frequently chastised by senior cult members if they fail to master these arts. A substantial number of novices are excommunicated from the cult at this stage. This sometimes happens at a formal ceremony called the *Oral* (an obvious contraction of the term oracle).

"Unique among the Physiologists is a sacred law which requires them to divide all animals into two groups, *Experimental* and *Control*. This is reminiscent of the Old Testament sacrificial rite where two identical goats were chosen, 'one for the Lord and the other for scapegoat.' After differing treatment, both goats are killed. Occasionally, a sacrificer will forget the sacred law of division and in the darkened room where he describes his rituals someone will intone, 'Where are your Controls?'

"The modern practitioners of 'zoocide' wear simple robes usually of white (for purity) or light green (for plant fertility?). They generally have a procedure of replacing robes that have been soiled in the sacrifice by fresh robes that have been ritually purified.

"After the sacrifices have been made and the blood and entrails have been removed (sometimes only a single organ is taken), the remains of the animal body are made into a burnt offering. This ceremony takes place in a subterranean room in a large fiery chamber called an *Incinerator*. Enigmatically, the burnt offering is made by an individual who is not a full member of the cult and often appears to do penance by sweeping the floor and removing trash. On rare occasions, the animals are not consumed by the sacred fire, but are eaten. This appears never to be the case when dogs, cats, or rodents are sacrificed, but occurs most often when the ceremonial animals are crustaceans such as lobsters.

"The Physiologists do not feel that their sacrifices are effective until they have been broadcast in great detail to the maximum possible number of fellow cultists. For this reason they print large numbers of religious tracts called *Journals* and undertake long and arduous trips to Laboratories around the world to describe their animal killings. The importance of informing others of the sacrifices is stressed in the phrase, 'Those of ye who do not publish, shall surely perish,' which is taught to novitiates during their training period.

"Some members of the cult practice the rite of withdrawing blood from living animals. Indeed, some even withdraw blood from living humans. Although human sacrifice is never practiced, the act of human bloodletting is quite common. This is thought to be a carryover from the vampire followers who were still flourishing in Eastern Europe at the time that Physiology arose. Indeed, the hollow tubes used to suck up the blood are probably modeled after the incisors of the vampire bat.

"The purpose of all of these sacrifices is to ward off disease, illness, and old age both for members of the cult and for the wider society around them. This is made clear from the frequent association of Laboratories with Hospitals. It is made even clearer from an annual rite in which each cultist writes a sacred confession and sends it to the capital of his country. The confession first begins by a terrifying description of the scourges and afflictions that the sacrifices are intended to cure. It then goes on to a section called *Background* where a rationale is presented as to how the sacrifices will alleviate the terrible conditions that have just been described. Then follows a long passage called *Progress* where the sacrificer humbly laments his failure to have already achieved his goal. This is always done in guarded ritual language lest any but the cognoscenti get their hands on the document. Finally, the confession ends with a plea for the entire nation to make 'a money sacrifice' to provide the cultist with the wherewithal to buy more sacrificial animals and portent machines in order to continue his efforts to restore the right relation of man to his sacred order."

Grant Us This Day Our Daily Bread

On occasion, when observing a talented scientist or scholar exhausting himself over the minutiae of a grant request, one drifts into utopian visions of a world in which well-supported seekers of truth pursue their goals in tranquillity. But reality always shatters this Panglossian image, and we must return to "cultivate our gardens." Sometimes, however, the encounter with the scholar-businessman sends one wandering into historical inquiry about our great predecessors.

The leading natural philosopher of antiquity was undoubtedly Aristotle. In his early days he supported himself by his teaching. This was followed by marriage to the sister of the wealthy Hermias, autocrat of the city-state of Atarneus. Aristotle then moved into the top teaching job in the country, tutor to Alexander in the court of Philip of Macedon. After his student went off to conquer the world, the universalist returned to Athens to establish his school, the Lyceum, which was the world's first major research institute of biology and natural science.

As to the financing of this effort, Athenaeus reports that Alexander the Great provided a grant of 800 talents (about $8 million in current terms) for the equipment and research. The conqueror also instructed his gamekeepers, gardeners, and fishermen to furnish Aristotle with zoological and botanical material and sent off research expeditions to provide information and specimens for the Lyceum.

The Athenian academy seems a vision of paradise: no grant requests, no peer review, no annual reports, no auditors. But even paradise has its snakes. Aristotle was forced to defend Alexander against the citizens of Athens who resented their loss of independence and the lordship of the Macedonian "barbarian." At the same time the philosopher argued with Alexander over excesses of the monarch. When his patron died and the Athenians declared their freedom, Aristotle was forced to flee the city. A few months later he died in exile. The great Lyceum came to an end, and for one thousand years a curtain of darkness fell over the pursuit of science.

Coming into the 17th century, we ponder the fate of René Descartes, who can be counted among the founders of modern science, mathematics, and philosophy. That savant's career was divided between periods of swashbuckling military activity and years of calm study and contemplation. (With some reluctance I point out to any teenage readers that his best thought was carried out while lying in bed late in the morning.) When he finally achieved the fully deserved recognition, he was called to the court of Queen Christina as official scientist-philosopher of Sweden.

Alas, however, his teaching and administrative responsibilities appear to have "done him in." He was unable to lie abed at the start of the day because his patroness desired his presence as a tutor at five o'clock each morning in one of the drafty halls of the palace. To compensate, he tried to arrange for an afternoon nap, but the duties of establishing a Royal Academy of Sciences intervened. The combination of the Swedish winter and the strong-willed young queen wore him down. Descartes contracted a respiratory disease and shortly thereafter he expired, a victim of his own success.

The story of Antoine Laurent Lavoisier is perhaps the saddest tale of all. The founder of modern chemistry earned his keep as a full titular member of the Ferme Générale, the main tax-collecting agency of Royalist France. His public service was altogether admirable and was of inestimable value to his country. However, all of these virtues were disregarded when a new administration came into office. After the revolutionaries became the establishment, there was little sympathy for former tax collectors. In November of 1792 the Revolutionary Convention ordered the arrest of all the former members of the Ferme Générale. On May 8, 1794, after a trial lasting half a day, Lavoisier and 27 others were condemned to death and guillotined. There are reports that the dedicated scientist asked for a few days to complete the write-up of his experimental work and that this request was denied. (At least we know that there is a cutoff to the writing of progress reports.) Lavoisier must be the first investigator in history to have lost his life for accepting support from the wrong governmental agency.

Who in 1870 (or today) would award a research grant to a country doctor, untrained in research, unpublished, and working to maintain a small-town practice? Robert Koch was an ordinary dispenser of medicaments in the East Prussian village of Wollstein when his wife Emmy awarded him what must be ranked as one of the highest yield research grants in history, a microscope, as his 28th birthday present. That gift set off a six-year project that led to understanding of the nature of anthrax and the formulation of Koch's postulates, which have been the cornerstone of medical microbiology.

Who supported those six years of intensive effort? First, Koch, who drove himself in endless hours spent in his tiny laboratory in the back of his office; second, Frau Koch, who endured years of loneliness and financial privation from the time he took away from his practice; and third, the sick of Wollstein, who competed for the doctor's attention with ideas that seized and transformed this man.

The story of Koch has a happier ending than some we have been relating. For when his discoveries were recognized, he was appointed Extraordinary Associate of the Imperial Health Office and given a laboratory and assistants to pursue his work. In this setting, he was to carry out many more years of the development of the germ theory of disease.

All of these random musings do not have a clear and well-defined conclusion. The support of intellectual effort has always been a capricious activity of society. To make matters even cloudier, it is not easy to see a very direct correlation between support and productivity. One thinks of Albert Einstein as clerk in the patent office or Gregor Mendel administering his monastery. What is different nowadays is that the high cost of equipment, services, and supplies makes it increasingly difficult for the loner to compete. The bureaucracy associated with the granting of funds makes it ever harder for radical ideas to be tested.

However, the human spirit is never totally overwhelmed, and still there appear to be alternatives for those daring enough to pursue them. I happily report that there is at least one present-day scientist who has been laboring away at his home laboratory for many years now and who has, I believe, made one of the major contributions of our century. Such individuals are not necessarily obvious to their fellow workers. History will one day report on those of our contemporaries who did it their way and succeeded.

Change Four Sparkplugs and Take Two Aspirin

Herbert Spencer, the great encyclopedic mind of the 19th century, was a dedicated optimist, thoroughly convinced of human progress. His first book, published in 1850, bore the ponderous title *Social Statics, or the Conditions essential to Human Happiness specified, and the first of them developed*. In this work he discussed the subdivision of labor which goes on within a human society as it advances, and pointed out that in the beginning every man was simultaneously a warrior, hunter, fisherman, builder, agriculturist, and toolmaker. The next step was the segregation of social units into classes such as warriors, priests, and slaves. This was followed by specialization among the laborers, and social development proceeded with the appearance of highly skilled, well-trained artisans. The advocate of progress concluded that further gain would be accompanied by increased subdivision into even more narrowly defined functions.

Lying in a pool of grease on the garage floor, helping install a new transmission, gave me powerful impetus to muse on Mr. Spencer's philosophic system. Each bit of rust and road dirt dropping in my face clued me that things had not quite turned out as predicted. The "law of individuation" was not being fulfilled by my straining to hold up the back end of the gear box as my son wrestled the front end into the clutch plate fitting. Surely the 19th-century savant had missed some aspect of our age; or perhaps we were in a retrogressive period where despecialization was overtaking us all. These *pensées* were interrupted by a shout of joy as the shaft snapped into place. We were ready for the bolts, and thoughts of *synthetic philosophy* disappeared in a pile of wrenches, wires, and assorted hardware.

Contemplation of Herbert Spencer was reawakened as months later I sat chatting with Thomas Ferguson, founder and editor of *Medical Self-Care* magazine. For within the books reviewed in this journal lay the seeds of a new trend, do-it-yourself medicine. Such a possibility warrants a reexamination of the entire do-it-yourself movement, the antithesis of Spencerian doctrine. Philosophically, these ideas go back to Ralph Waldo Emerson's 1841 essay *Self-Reliance*, where we read: "A sturdy lad who in turn tries all the professions, who teams it, farms it, peddles, keeps a school, preaches, edits, goes to Congress, buys a township and so forth and always like a cat falls on his feet, is worth a hundred of these city dolls." Emerson was intellectualizing a tradition from a frontier society which required each man to be all things, while Spencer in the more mature society of England could build theories on the specialization of function. For the more classical among us, the idea of do-it-yourself may go back all the way to the ideal of Renaissance man, one skilled and well versed in nearly all the arts and sciences.

However, interesting though these philosophical theories may be, I believe that the thrust of the modern do-it-yourself movement is largely economic rather than ideological. As specialization increased in this century, the hourly wages of trained workers also increased, so that the cost of hiring experts for each task became too much for the average family. We began to see a large number of books and articles on build-it-yourself, home auto repair, plumbing, electricity, TV servicing, and a variety of other crafts. Individuals, by acquiring generalized skills, could considerably ease their financial burdens while at the same time enjoying the

Emersonian sense of reliance. Of course, the price of this solution came in the form of a number of jobs poorly done by inept amateurs.

The income tax structure, which places a high toll on every transfer of money, has acted to further despecialization. A physician in the 50% bracket must earn $100 to pay the plumber $50. The plumber, also in the 50% bracket, must charge $50 because he ends up with only $25 to pay his bills. The result is that the physician unclogs his own drain. However, the reverse economics also applies, and we may be at the beginning of the period where the plumber will attempt to unclog his own gastrointestinal or urogenital system. All of which takes us back to *Medical Self-Care*.

The magazine under discussion is not designed to replace the physician. It is rather the voice of holistic medicine, the concept that mind and body are one and each individual bears responsibility for his own health. That responsibility includes a life-style designed to promote well-being, a certain amount of self-diagnosis as well as knowing when to seek out a physician. The journal describes itself as "a medical *Popular Mechanics*, a *Whole Earth Catalog* of the best medical books, tools, and resources, a *Consumer Reports* focusing on health care."

The thrust of the self-care concept is perhaps best seen in some of the titles of the books reviewed: *Take Care of Yourself, A Consumer's Guide to Medical Care, How to Be Your Own Doctor (Sometimes)*, and *Personal Health Appraisal*. Although there is a certain counterculture flavor to the approach, the authors are mainly physicians, as is the editor. The movement, while certainly not at the center of the medical establishment, cannot be described as being totally outside it.

The concept of medical self-care opposes a tradition that has developed throughout the twentieth century. The rigorous licensing system for physicians and the prescription system designed to keep all effective reagents under tight control are but examples of a public policy based on the postulate that medicine is too complex, technical, and specialized to be entrusted to any but approved experts. The counterargument being introduced is that health is, in large part, a matter of right living, right thinking, and an individual's care of his own body. The adherents do not deny the role of the physician but transfer it to one of repair jobs when necessary. The new view stresses protective service and maintenance rather than fixing breakdowns.

There is no telling where this will lead. Many trends are apparent in modern medicine, and no one, least of all this writer, is prophetic enough to know which trend will predominate. The idea of do-it-yourself medicine does surely conjure up some strange images. The home wine enthusiasts might go into antibiotic fermentations. An artificial kidney kit is surely as challenging as a harpsichord. And building a diagnostic x-ray machine might be just the thing for the electronics buff. Doubtless, all of this would have come as a great surprise to Herbert Spencer, but then the modern world has been tough on a lot of 19th-century theories.

On Swallowing the Surgeon

A book or article on an unfamiliar subject often sets the mind off in new and surprising directions. A copy of *Symposium of Microsurgery* (A. I. Daniller and B. Strauch, Eds., The C. V. Mosby Co., St. Louis, 1976) thus evoked long-resting memories of a 1966 science fiction movie, *Fantastic Voyage*. In that saga, a small submarine with a crew of five was miniaturized by some rather mysterious physical process to the size of a bacterium. The surgical team was then injected intravenously and sent on a mission through the bloodstream to remove a lesion from the brain of Dr. Benes, an incomparable scientist from "the other side" with secrets of monumental importance. Of particular fascination to me was the description of the terrors of aortic turbulence when viewed from the inside of an underseas craft a few microns in length.

Several years before the fictional account of medical miniaturization, Nobel laureate Richard P. Feynman had raised this very issue from the scientific perspective in an imaginative and entertaining essay entitled "There's Plenty of Room at the Bottom" (reprinted in *Miniaturization*, H. D. Gilbert, Ed., Reinhold, New York, 1961). He started out by posing the question, "Why cannot we write the entire 24 volumes of *Encyclopaedia Britannica* on the head of a pin?"

The answer, stated with great optimism, is that such a task is theoretically possible and requires only technological ingenuity. At this scale all the extant books in the world could be transcribed into a small pamphlet; librarians take notice.

Feynman's analysis of smallness then went on to a discussion of computers and concluded that great reduction was possible. Since the writing of his essay in 1959, there has been a size decrease of orders of magnitude in this hardware, as witnessed by a hand-held calculator sitting in front of me that can far outperform the large desk-top model that I used 20 years ago.

The next item was miniaturizing machines, and the author passed on a suggestion of a friend of his, Arthur Hibbs, that very small mechanical devices could be used therapeutically. "He [Hibbs] says that, although it is a very wild idea, it would be interesting in surgery if you could swallow the surgeon. You put the mechanical surgeon inside the blood vessel and it goes into the heart and 'looks' around. (Of course the information has to be fed out.) It finds out which valve is the faulty one and takes a little knife and slices it out." While cardiac surgeons might be amused by this description of valve repair, the general idea is clear: tiny computer-controlled devices could work from inside the body much as the submarine functioned in *Fantastic Voyage*.

The *Symposium of Microsurgery* gives us occasion to examine how far we have come in miniaturizing surgical hardware. The discipline is defined in the first article as the art of operating with the aid of a microscope. The technique is used for the union of blood vessels in the size range of 1 mm and in nerve repair. The smallest pieces of equipment reported are suturing needles formed by electroplating the ends of single strands of nylon thread in the size range of 7 to 15 microns diameter (about one hundredth of a millimeter). Needles of 70 microns are generally used along with jewelers' forceps, clamps 8 mm long, and similar microinstruments. At present the instruments are manually controlled under microscopic observation, and the greatest dexterity is required.

Moving from human to animal studies we note that for many years protozoologists have been performing microsurgery on single amoebae. One of the most impressive feats is a nuclear transplant in which the nucleus from one protozoan is inserted into a second cell which had previously undergone a nucleo-ectomy. Such experiments, carried out with micromanipulators, allow for a study of the relative function of nucleus and cytoplasm in large unicellular forms.

Another kind of miniaturization was shown to me by an entomology student who makes tuned circuits on very tiny discs that are glued to the backs of bees. Appropriate circuitry allows the experimenter to record the goings and comings of individual insects as their antennae pass his antenna at the entrance to the nest. In Feynman's terms all of these examples are only the beginning of size reduction, and there is still plenty of room at the bottom.

However, there are limits to fabricating small machines which relate to two fundamental properties of matter: atomicity and molecular motion. In 1827, the English botanist Robert Brown reported that microscopic pollen grains in aqueous suspension underwent a continuous zigzag motion. He showed that this type of movement is characteristic of all very small suspended particles, both living and nonliving, and the designation "Brownian motion" honors his contribution. The cause of these irregular particle trajectories remained an enigma until near the end of the century. It was then postulated that the phenomenon was due to collision of the particles with fast-moving solvent molecules. Einstein finally worked out a theoretical explanation of Brown's observations, and subsequent experiments confirmed this theory and led to a determination of molecular sizes.

Brownian motion is only one example of an intrinsic feature of molecular structures known as random thermal motion. Fluctuations are also seen in electronic circuit noise and light scattering. At the atomic level all particles are ceaselessly moving about and colliding. Temperature is, of course, a measure of this motion. A consequence of this randomness is a limitation on the smallest possible dimensions of working apparatus. Below a certain size the thermal motion makes it impossible for a functioning piece of hardware to carry out a precise job.

If we are prevented by a fundamental law of physics from building apparatus below a certain size, we must ask how a living cell manages with much smaller working parts. That question was raised by Erwin Schrödinger when he examined biology from a physicist's perspective. He pointed out that most physical laws achieve precision because they are averaged over a large number of molecular events, while cellular processes like gene replication involve only a small number of molecules. Precision in the latter example comes from quantum mechanical transitions of covalently bonded structures; biologically miniaturized machines must thus be single molecules. There are two types of precision: order from disorder in conventional machinery containing a large number of atoms, and order from order in such gadgets as enzymes and membrane-bound energy transducers.

We are tempted to ask, can microsurgery be done with molecular machines? The answer is yes, but in raising this question we have subtly moved from surgery to pharmacology. At an abstract degree of miniaturization, this move defines the line between surgery and internal medicine. In the former we work with supercellular structures such as tissues to manipulate the system to a healthier state. In the latter case we chemically attack problems at the subcellular domain to improve that state of the patient's health. It is strange that medical specialties divide at such a fundamental level.

Will ordinary miniaturization ever get good enough so that we can swallow the surgeon? It is hard to predict what future technological problems might arise. There is certainly a wide-open and exciting field of biomedical instrumentation of extremely small components, and we are very far from any theoretical limits. What we may certainly conclude is that in Feynman's terms we have been swallowing the internist ever since systemic drugs first came into being.

Deep Purple

In the dense waters of the Dead Sea and Great Salt Lake, in drying tidal basins and in pickling brines, are found strange, atypical life forms capable of living in such extreme environments. Among these oddities are the halophiles (salt lovers), bacteria that not only survive but even require abnormally high concentrations of sodium chloride. The story of these organisms is beginning to unfold as a fable with a moral worthy of Aesop: never be too sure that you know where future research will be heading.

Our narrative begins in 1880 with the publication of a U.S. Government monograph entitled "On the Nature of the Peculiar Reddening of Salted Cod Fish." This study, early in the days of bacteriology, revealed the microbial nature of the coloring agent found in high-salt environments, thus simultaneously explaining the hue of evaporating salt ponds and the tint of rotting fish. The detailed nature of these red bacteria was elaborated almost 40 years later in a German publication on the contaminants of dry cod.

The scenario now shifts to the area of the Dead Sea where a graduate student from Hebrew University was proving that those waters were not lifeless and was indeed characterizing the inhabitants of that brine as well as other high-salt organisms. The Ph.D. thesis of B. Volcani (1940) conferred the name *Halobacterium halobium* on the particular variety of red salt bacteria that are the central characters in our present-day drama.

For the next 30 years the halophiles were of minor importance in the science of bacteriology, studied largely by food technologists and cell physiologists interested in biota able to survive in high osmotic pressure surroundings. During this period an average of about two papers a year on *Halobacterium halobium* appeared in *Biological Abstracts*, reflecting the relatively low level of interest in these entities. Then A. D. Brown, an Australian researcher on osmophilic organisms, made the suggestion that, because of certain peculiar features, the cell envelope of *Halobacterium halobium* could serve as a model system in the study of plasma membranes. This proposal was taken up by a membrane ultrastructure group in San Francisco under the direction of Walther Stoeckenius. Other laboratories subsequently joined in. Several quite unexpected results emerged in rapid order.

First, it was found that under oxygen deprivation these bacteria respond by producing a different kind of cell membrane, purple in color and containing the pigment retinal, a surprisingly close relative of one of the light-absorbing molecules of the human eye. Second, it was ascertained that these patches of purple can absorb light and convert that energy into making ATP, the universal molecule in biological power transfer. The third finding was the mechanism of this photosynthesis of ATP. Under the action of light, the specialized membrane moves hydrogen ions from the interior of the cell into the surrounding medium. The energy released in the backflow of these ions is coupled to the production of ATP.

With these findings, interest in the salt-loving bacteria grew rapidly. In 1973 and 1974, *Biological Abstracts* reported six papers per year on *Halobacterium halobium*. In 1975, there were 14 publications, and in 1976, the number had grown to 33. For the last complete abstracts, 1977, about 50 articles have been added to the literature.

To appreciate why these results have been of such keen interest to the biochemistry community, we must unravel another strand in our story, the problem of how higher animals process their energy into a biologically useful form. Since the late 1930s, it has been known that the key to this puzzle lay in molecules of ATP, since almost all energy passes through this intermediate form on its way toward utilization in the body. The beating heart, the active cerebrum, the flexing biceps, all are driven by ATP power. Further studies showed that most of this energy conversion takes place in the mitochondria, cellular organelles responsible for the controlled combustion of sugar and oxygen to carbon dioxide and water.

Many theories were proposed and bioenergeticists were rather divided as to the mechanism whereby the energy of oxidation became converted to the chemical potential of ATP. Around 1960, Peter Mitchell in England proposed that oxidation was accompanied by the energy requiring transport of hydrogen ions across the inner mitochondrial membrane. He further postulated that the return flow of these ions was coupled to chemical reactions yielding ATP.

The reason for the excitement attending the *Halobacterium* studies is now apparent. Both the purple membrane and the inner mitochondrial membrane function as energy-driven hydrogen ion pumps. However, while the mitochondrial system is enormously complex, involving a large number of proteins and chemical intermediates, the bacterial system is far simpler. The purple membrane contains only one species of protein of rather low molecular weight. Thus arose the hope that the mitochondrial bioenergetics problems could be illuminated by the studies of the newly discovered purple protein. These expectations were furthered in an experiment in which the *Halobacterium* hydrogen ion pump was combined with the mitochondrial ATP synthesizing apparatus in an artificial system, and the former exposed to light was able to drive the chemical reactions of the latter.

Like all great contemporary historical sagas, ours must remain unfinished as the passage of time adds chapter and verse to the unfolding story. We have come far enough to realize that *Halobacterium halobium* will play a major role in elucidating bioenergetic mechanisms. Few now doubt the potential importance of this understanding in medicine and agriculture. We must also look backward and realize that there is no way that the immense importance of this organism as an experimental tool could ever have been realized from the early studies on the discoloration of codfish or any of the research prior to 1970. The tale of *Halobacterium* thus demonstrates our moral: the unpredictability of the advance of science.

We now move from fable to sermon. Present-day funding of research is growing more and more programmatic, closing off opportunities to explore byways and biological oddities. It becomes increasingly difficult to raise the money to investigate branches of science which promise no short-term practical rewards. Yet, I hope our salty tale has convinced you that there is no way of predicting where the payoffs will come from. More familiar examples are the discovery of antibiotics starting with Fleming's study of mold contaminants on bacterial cultures, and the knowledge of x-rays originating in Roentgen's investigations of electrical spark discharge in gases.

Doubtless projects on why salt cod turns red or whether there is life in the Dead Sea would be candidates for Senator Proxmire's Golden Fleece award. Yet such studies have unquestionably contributed to the understanding of one of the real enigmas of bioenergetics, with much promise for the future. Dr. Stoeckenius, who pioneered the role of the purple patches, doubts that he, as a mammalian membrane microscopist, could have gotten federal grant support to study salt bacteria. What would have been difficult in 1969 is even less likely by today's standards. *Halobacterium halobium* stands as a symbol that we never know when the pauper-clad may become the purple-robed. Do you read me, Senator?

Shrinks

In some circles the slang word "shrink" is now universally employed as a flip synonym for "psychiatrist." The Dictionary of American Slang records the earliest usage occurring in 1957 when the epithet appeared as "headshrinker." The etymological transition from psychiatrist to headshrinker to shrink probably has reference to the mores and customs of the Jivaro Indians of the upper Amazon basin. This primitive tribe has gained worldwide recognition because of its unusual custom of cutting off an enemy's head, shrinking it to the size of a fist, and then dancing around it in a joyous ritual. Since the only readily apparent similarity between the Jivaro warrior and the psychiatrist is a common focus on the head as a site of power and mystery, the association seems a very tenuous one at best. The word linkage probably reflects an underlying fear of anyone who may gain control of the contents of our heads.

Since psychiatrists are much more common than headhunters among my usual acquaintances, I was pleased some time ago to meet Lewis Cotlow, a man who on three occasions has lived among these South American Indians and tried to comprehend the hows and whys of a social system so very different from our own. His book *Amazon Head-Hunters* provides a fascinating account of these experiences.

Before venturing into the anthropological and psychological aspects of this matter it is instructive to examine the surgical and biophysical protocol. Following battle, a victim is placed on his back and decapitated from the front with a sharp lance. The final step in the operation is the insertion of the tool between two vertebrae low on the neck. The disembodied head is then carried on palm strips that have been run through the mouth-neck channel. On the way back to their home village the warriors stop and carry out the miniaturization procedure.

An incision is made from the back of the neck to the top of the head and all bones are removed along with the brains. The eyelids are sewn shut from the inside while the lips are clamped with wood pins and fibers. Protein denaturation is assured by two hours of boiling the head in a mixture of water and vine juices. After this preliminary treatment the original incision is closed and hot stones are placed in the sack formed by the stitching. Upon cooling, the rocks are replaced by fresh hot ones, and the process in repeated. The features are carefully shaped during the shrinkage and eyelashes and eyebrows are plucked to keep all parts to scale. After sufficient drying and size reduction the residual head, now known as a *tsantsa*, is hung over a smoky fire for the final processing. This is an impressive if grisly bit of biotechnology carried out by these Stone Age people.

Having outlined how they do it, we are left with the more serious problem of why. Here Cotlow suggests that the answers lie deep within the metaphysical view of reality held by the Jivaros, which is surprisingly causal and theoretical. They believe that all events have underlying reasons; nothing of significance happens by chance. If, for example, a tree falls on someone, it is the duty of the witch doctor to go into a narcotic-induced trance and determine the cause. The most frequent source of trouble is believed to be a spell cast or demons unleashed by some member of a neighboring tribe. The unhappy soul of the departed calls out for vengeance, which requires that the enemy be transformed into a tsantsa.

Each revenge must be re-revenged, so that a constant state of headhunting exists.

Lest we simply dismiss the activities of the Jivaros as the senseless acts of primitives, a closer look at the society is in order. Within the villages families lead a responsible, happy existence. Husbands and wives show love and affection, and children and parents exhibit mutual feelings of attachment. Fathers and mothers spend much of their time teaching children the skills and lore of their group. Life is hard in the jungle, and these Indians have learned to survive on hunting, fishing, and farming, each of which presents major difficulties.

Cotlow interviewed the great chief Utitcaja, who had taken 58 heads. "He felt as proud of having killed his family's enemies as the Crusader did at killing infidels, the enemies of his God and his faith. He was a happy man because he had done so much good in the world—his own world, but above all the spiritual world of his ancestors." The tsantsa then is a sacred object necessary to the Jivaro because he lives in a world of demons, ghosts, spirits, and souls of unrevenged relatives. He must deal with this world using the knowledge that has been faithfully transmitted by his parents.

Thus, in a strange way, the tsantsa religion and the practice of psychiatry may have a relation that goes beyond the original ill-intended use of "shrink." Many of us in contemporary Western culture also live in personal worlds of demons, ghosts of the past, and shades of departed relatives. We seek our surcease in the ways provided by our society. As Cotlow has noted:

> Chumbika shrinks a head he has cut off
> John Doe goes to church
> Mary Roe consults a psychiatrist
> They seem to have vast differences.
> But the motivation is the same in each case—
> peace of soul.

A second curious similarity appears between the Jivaro and Freudian philosophical foundations. Both depend on a belief in a certain kind of strict causality; little can be ascribed to chance. Thus the notion of a "Freudian slip" is an assertion that an event which might have been regarded as a meaningless accident is in fact causal in nature, the result of past happenings which may have occurred many years ago. Much of therapy is directed at identifying the causes by an effort to recover the past. The relation between the two world views exists in the shared importance of determining precedents, but the range of possible causes and the domain of permissible responses are extraordinarily different.

In these terms, comparative anthropology comes into its own and we can look more sympathetically at each culture and age in its own terms. We don't have to like what we see, but it is possible to investigate what those disturbing activities are doing for their practitioners. One observation is of the incredible differences and strange similarities between societies, not so much in physical culture, but in the way they structure their worlds with metaphysical conceptions. This is well known to anthropologists, but comes into sharper focus when we expand our minds by following the implications of the word "shrink."

The Jivaros' world is no less real to them than ours is to us. When faced with the problem of which is the more correct view, we, including our psychiatrists, are very sure of the answer, but, as Cotlow discovered, so are the Indians.

PART V

The Crazy, Hairy, Naked Ape—REFLECTIONS ON SOCIAL ANIMALS

The Crazy, Hairy, Naked Ape

For the past several years we seem to have been somewhat hung up on our primate taxonomic status and the behavioral antics of our simian relatives. The individuals most directly involved in putting us in this bind are, of course, Charles Darwin and Thomas Huxley. William Irvine has reminded us of this point by calling his combined biography of these two *Apes, Angels and Victorians*. In response to this concern over our apeness, it behooves us to go back a bit and examine the genealogy of the group in more detail.

The primates are an order of mammals in the systematic scheme and are split into suborders, two of which contain relatively primitive animals such as tree shrews, lemurs, tarsiers, and the like, and a remaining order, the Anthropoidea, containing monkeys, apes, and you and me. However liberal we may be, we are, at all intellectual levels, somewhat embarrassed by our earlier relatives, a point that is never lost by the propagandists against the theory of evolution. Eugene O'Neill gave us the reaction of the totally uneducated Yank who perceived he had been called "a Hairy Ape." The pathetic Yank, with doubts about his own humanity fostered by rejection by his fellow humans, cannot cope with being seen in this context. We view him somehow as the missing link, part man, part ape, and belonging nowhere. He is a reminder of the apeness in each of us. It was easier when we were a little lower than the angels. Now that we are but a little higher than our anthropoid relatives, it is slightly more difficult to look into the mirror in the morning, particularly for those of us who are cutting off our facial hair.

The order Anthropoidea mentioned above includes four families: the primitive New World monkeys and marmosets and the more advanced Old World monkeys and apes. The anthropoid apes, or Pongidae, are the primates closest to man. It was while comparing humans to the Pongidae that Desmond Morris first referred to man as "The Naked Ape," meaning a primate with far less body hair than any of his close relatives. The primary anatomical features of the entire group of apes are the absence of a tail, the more or less upright posture, and the high degree of brain development. One wonders how it was that George Bernard Shaw's uncle was unaware that apes lacked tails. For Shaw writes: "What he repudiated was cousinship with the ape, and the implied suspicion of a rudimentary tail, because it was offensive to his sense of his own dignity, and because he thought that apes were ridiculous, and tails diabolical when associated with the erect posture."

It was embarrassing enough when our relationship to the apes was based on anatomical and evolutionary grounds; the chagrin has become more acute with the development of contemporary behavioral biology and the subsequent realization of the psychological similarities between us and our hirsute relatives. This aspect most disturbed us when Desmond Morris discussed our origins, sex habits, rearing patterns, exploration behavior, fighting, feeding, and comfort.

The anthropoid apes are represented in a modern zoo by gorillas, chimpanzees, gibbons, and orangutans. A number of related species are known only from fossil evidence. Among these extinct forms evolutionary theorists assume the existence of the most recent ancestors of the now ubiquitous *Homo sapiens*. The descent of man has not been traced in complete detail. Fossil material is limited and is likely to remain so for the foreseeable future, so we are not likely soon to develop our evolutionary genealogy in complete detail. In any case the features that most distinguish man from ape are psychological in nature and these are even more difficult to reconstruct from fragments of bones and teeth.

Albert Szent-Györgyi had in mind this difference in mental capacity between humans and other primates when he used the epithet "The Crazy Ape" as the title of a book. He reasoned that since the development of science and culture have shown *Homo sapiens* to be an extremely intelligent primate, the failures in politics and social order must be due to "the machinations of present-day men who often appear to act more like crazy apes than sane human beings." What is interesting in the above context is that even for the highly sophisticated thinker Dr. Szent-Györgyi, the term "ape" is an epithet in the same sense that it was taken by the crude, uneducated character Yank in Eugene O'Neill's play. Szent-Györgyi was angry with his fellow humans and he resorted to name calling to add emphasis to his point.

The two people who have probably done the most for the apes in a public relations sense are Edgar Rice Burroughs and Jane Goodall. Burroughs, the novelist, created Tarzan, the son of an English nobleman who was raised and nurtured by a tribe of great apes in Africa. Tarzan became a very popular hero and the apes were his constant allies in his confrontations with "the bad guys." The villains were most often *Homo sapiens*, so the jungle animals came out clearly ahead. Jane Goodall is a scholar whose studies on the chimpanzees of East Africa are major contributions to the behavioral biology of primates. Her movies and articles in *National Geographic Magazine* have attracted attention among a much wider public than the animal ethologists. Chimpanzees emerge from these studies as individuals with personalities, rather likable on the whole. "Ape" becomes much less of a derogatory term after one becomes familiar with Dr. Goodall's observations.

Darwin's theory of human evolution caused a great perturbation in man's self-image. For thousands of years Western man had envisioned himself as existing apart from nature. Evolutionary thought not only revealed man's primate status but placed him right in the middle of the natural world. For the last hundred or so years, that concept has been working its way from the centers of learning through society at large. It is a very painful notion. To be suddenly removed as a very special child of the Creator and placed in the zoo with all the other animals is a traumatic experience. Human society has not recovered from the shock. The literary responses to our own apeness that we have been discussing reveal the ambivalence at many levels. The British clerics, such as Bishop Wilberforce, who so violently opposed Darwin, must have sensed the deep damage to the existing order of things that would follow from evolutionary theory. Though we tend to berate them as intellectual obstructionists we should appreciate the validity of their concern with the profound effects of change in the human self-image. Thus in debate the Bishop turned to Huxley and "begged to know, was it through his grandfather or grandmother that he claimed his descent from a monkey?"

If we, as a society, are still uneasy about our primate status, it is an understandable malaise. Our position has eroded over the past few hundred years from being the center of the universe to being one more species on a small planet orbiting a medium-sized star in one galaxy out of the multitude of galaxies that exist in the universe. It is from this humble starting point that we must begin to recreate love, beauty, and truth. It is a truly gargantuan job that leaves us little time to monkey around and certainly does not permit us simply to ape the intellectual attitudes of our predecessors.

Social Implications of a
Biological Principle

A mong students of "biotic interactions" there has developed a controversial law or principle that is of more than
academic interest, since it has social implications of the most
far-reaching scope. The law has had many statements and
many names and has become best known as The Principle
of Competitive Exclusion, or Gause's Hypothesis, after the
Soviet biologist G. F. Gause, whose experimental studies in
the area appeared in a 1934 monograph, *The Struggle for
Existence* (Williams & Wilkins Company, Baltimore, Md.). The
principle as set forth in the *Encyclopaedia Britannica*, 15th
edition, states that "Populations of two species cannot persist
together for a very long time in the same community when
both compete for and are limited by a common resource."
The social problem to which this principle relates is that of
two culturally isolated, noninterbreeding populations that
occupy the same country. We will consider Northern Ireland
as the archetypal example of the topic to be discussed.

Before we can focus on the social problem we must go
back a bit into evolutionary and population biology to understand the implications of the Gause Hypothesis. Biologists usually define a species in terms of an interbreeding
population. If such a population is isolated into two subgroups
that do not interbreed because of geographical barriers, then

each group experiences different environmental pressures and different genetic drifts. Eventually the descendants of the two original populations, many, many generations later, will be sufficiently speciated so that they will not interbreed. Two species will have arisen from a single initial species. Between the original species and the final two species there will be an intermediate stage at which the two groups will show morphological differences but will still interbreed. At this stage the two groups are subspecies or races. The same factors operate in human populations and it must be argued that if any human subgroup were kept in complete reproductive isolation from the rest of mankind it would eventually become a new species. Because human sexuality is periodically stronger than cultural barriers, this type of speciation seems most unlikely for *Homo sapiens*.

In any case, the principle of competitive exclusion deals with two closely related species, for this closeness is implied in competition for a common resource. For example, two species of birds, one of which eats seeds exclusively and the other of which eats worms exclusively, are not limited by a common resource and are not in competition. Competition in the ecologists' sense involves occupying the same "niche," playing the same role in the community food network. Indeed, another statement of this principle is that no two species can occupy the same niche in a given community.

The consequence of the principle we have been discussing is that when two similar species find themselves in direct competition one of two things happens: 1) one species is eliminated by death or displacement; 2) one species modifies its behavior either adaptively or genetically so there is no longer competition for a common resource. The two species can then coexist.

Two noninterbreeding human populations in the same community present a close analogue to the Gause Hypothesis. While in nature we have only genetic speciation, in human interactions we have cultural speciation, where members of two groups in a community do not interbreed for cultural rather than for biological reasons. If the two groups are in competition for a limited resource such as land, food, jobs, or housing, we must assume that the principle of competitive exclusion will operate and we will observe a period of unpleasantness and strife that will result in one of the sequelae listed above. In human populations there is a third possibility: the cultural barriers are lowered and the groups merge to a single unit where intragroup competition rather than intergroup competition governs the subsequent behavior.

Examining Northern Ireland, we have an indigenous Roman Catholic Irish population and an intrusive Protestant English and Scottish population. The two groups have coexisted since the 17th century and the amount of interbreeding has been so limited that almost everyone falls into one of the two groups. With the rising tide of expectations following World War II the two groups came into obvious competition for limited resources. This competition was voiced in the complaints of discrimination in housing and employment that were made by the economically disadvantaged Catholic population. The result of this situation is a "struggle for survival" in the Darwinian sense. The struggle is indeed a costly one in terms of human lives and human values.

The question we must raise asks about the possibility of a solution that does not consider the principle of competitive exclusion. Must not statesmen take into account the deep-rooted biological construct in the search for a solution? The theory allows three solutions: 1) one group drives out or destroys the other; 2) two niches are created by partition or the creation of a highly class-oriented society with little resource competition; 3) cultural despeciation or the end to noninterbreeding. The theory suggests that any other solution will lead to continuing conflict.

All three solutions have occurred in the history of human societies. The Ugandan native population has recently driven all of the Indian community out of the country. In classic Hindu society with closed stratification, various noninterbreeding groups, castes, had different functional roles so that competition was avoided by the assignment of niches to various populations. Indeed, in some communities, caste was equivalent to functional role in the society. The State of Hawaii appears as an example of interbreeding eliminating the competition of cultural speciation, although the process is far from universally operative.

There are many countries in the world where two or more relatively noninterbreeding groups are in competition. The biological background suggests that these situations are unstable and the existence of political strife is widely noted. Examples are blacks and whites in the United States, and Flemish- and French-speaking peoples of Belgium. It is important to realize the relation of these social situations to the analogous cases in Gause's Hypothesis. Realistic political solutions must occur within the background of our biological nature and the underlying laws that govern the interactions of organisms. Where none of the biological solutions are acceptable to the populations involved in competition, we must anticipate continuing strife. This seems a harsh conclusion, but we cannot crawl outside of our biological skins to solve social problems.

A Controlled Social
Experiment

D r. Richard Clarke Cabot (1868-1939) was simultaneously
Professor of Clinical Medicine and of Social Ethics, "thus
filling twice as many chairs as it is possible for most mortals
to occupy at Harvard in our generation." He was, therefore,
in a unique position to propose and participate in the design
of the Cambridge-Somerville Youth Study, an effort to test
by controlled experiment the effects of counseling, friendship, and maximal use of social service agencies on the lives
of boys who were judged to have a high probability of being
predelinquent. Cabot believed that "someone should come
to know and to understand the man in so intimate and
friendly a way that he comes to a better understanding of
himself and a truer comprehension of the world he lives in."
However, the zeal of the Professor of Social Ethics was tem-
pered by the realism of the Professor of Clinical Medicine,
which led to an experimental design stressing the concept of
"controls." This is a rarity in social science where one does
not often have the opportunity to think in terms more usual
to the natural scientist. A total of 650 boys between the ages
of five and thirteen were chosen. From prior correlational
studies they constituted a cohort likely to have a high rate of
juvenile delinquency. They were grouped in 325 matched
pairs, each consisting of individuals of similar age, back-

ground, and behavior profile. By a random procedure, one
member of each pair was chosen as a control and the other
as a member of the treatment assemblage. "This latter group
was to receive all the aid that a resourceful counselor backed
by the Study, the school, and community agencies could pos-
sibly give" (*An Experiment in the Prevention of Delinquency*, by
Edwin Powers and Helen Witmer, Columbia University Press,
1951).

Treatment varied with the counselors, but in all cases they
visited the families and had frequent talks with their advisees.
Trips and recreational activities were arranged. More than
100 of the children received medical or psychiatric attention.
One quarter were sent to summer camps, and almost all were
brought into contact with such organizations as the Boy
Scouts and YMCA. Over half of the students were tutored
in reading and arithmetic. They, as well as their families,
were encouraged to attend church, and priests and ministers
were alerted to their problems. In short, all the social services
of the community were brought to bear to rescue these youths
from criminality and asocial behavior and to orient them to
healthy, productive lives.

The treatments began in 1939, and the average time in
the program was five years. A number of follow-up studies
have been made. The first was reported in the book by Powers and Witmer cited above. A second study by William
McCord and Joan McCord was published in 1949 under the
title *Origins of Crime* (Columbia University Press). The latest
analysis is "A Thirty-Year Follow-up of Treatment Effects"
by Joan McCord of Drexel University (*American Psychologist*,
33:284, 1978). Dr. McCord has managed to locate 480 men
from the 253 matched pairs who remained with the program. Lives have been traced through court records, mental
hospitals, alcoholic treatment center records, and vital statistics in Massachusetts. Responses to questionnaires were received from about half of both treatment and control groups.

Of those who responded to the questionnaire, 72 from the
treatment group had·a juvenile criminal record as compared

with 67 from the control group. Convictions for serious crimes as adults were 49 and 42 for treatment and control men, while 119 and 126 respectively had minor criminal records. Thus, when compared with respect to both juvenile and adult records, the groups were similar, although men from the treatment group were more likely to be recidivists (78% vs 67%). Comparisons of the two cohorts with respect to alcoholism, mental illness, or stress-related disease indicate that in those cases where differences are judged to have statistical significance, the treatment group is always somewhat worse off than the controls. Similarly, in job classification and job satisfaction, the control group seems to be better off than those who had the counseling.

When asked "In what ways (if any) was the Cambridge-Somerville project helpful to you?", two-thirds of the respondents in the treatment group believed that the project had improved their lives and put them on the right road. Many expressed fond recollections of their counselors.

In examining the Cambridge-Somerville study and Dr. McCord's analysis of the 30-year follow-up, it is important to note their special character because of the rarity and difficulty of truly controlled experiments in sociology and psychology. The dedication of Cabot, McCord, and a host of others presents us with "hard" information which must be taken into account in developing programs to deal with the problems of our society. And the news is not good, because we are brought face to face with inadequacies, uncertainties, and disproof of a number of cherished conjectures of generations of social workers and community planners.

The McCord report is a document of absolutely top priority that should be read by every social worker, psychologist, legislator, and public policy planner in the United States. Facts are surprisingly few in this area, and when we have them established with great methodological skill, they must be used to best advantage. As Dr. McCord states: "The objective evidence presents a disturbing picture. The program seems not only to have failed to prevent its clients from committing crimes—thus corroborating studies of other projects—but also to have produced negative side effects. . . . Overall, however, the message seems clear: intervention programs risk damaging the individuals they are designed to assist. . . .

"These findings may be taken by some as grounds for cessation of social action programs. I believe that would be a mistake. . . . We should, however, address the problem of potential damage through the use of pilot projects with mandatory evaluations."

To extract one additional message from the Cambridge-Somerville study, it is instructive to compare the procedures that accompany the introduction of a new drug with those followed in the introduction of new procedures in psychology or social psychology. In the former case, we first institute animal and tissue culture studies with treatment, no treatment, and placebo controls. If successful, we move to experiments on small groups of humans, again allowing for controls. Under the watchful eye of the Food and Drug Administration, larger-scale human studies precede the release of the agent for general use. After release, follow-up studies continually monitor the treatment for both efficacy and undesired side effects. In the domain of the mind we exercise little or no control and few controls. Each practitioner develops a treatment protocol and applies it without regulation and with only his own intuition and those of his colleagues to evaluate the efficacy and side effects. As the Cambridge-Somerville study shows, intuitions can be very misleading, and even the responses of the treatment group are of little objective value. The presence of controls in pharmacology allows the development of the treatment of choice; the absence of controls in all forms of behavioral psychotherapy eliminates on methodological grounds the possibility of scientific choice.

The question must then be asked: Why do we support behavioral therapeutic measures that do not meet the minimum criteria of scientific validity? Indeed, why do we support treatment protocols without "hard" knowledge of whether they are helpful or harmful? I suspect that when faced with the problem of people in need, we are unable to accept the humbling dictum that if we do not know what we are doing, it is better to do nothing than to do something. We assume that our intuitive responses are better than in-action. The studies that we have been discussing show the fallacy of this activism in the absence of methodologically secure knowledge.

Without empirical bases, social action programs become experiments rather than therapies. As such, they should be kept on a reasonably small scale and should contain built-in controls to assess their validity. These are not abstract issues; enormous efforts and resources are expended in the United States on social action programs. The Cambridge-Somerville study makes us confront the questions: Do we know what we are doing? And, if not, What are we going to do about it?

It would be somehow inappropriate to terminate this essay without expressing a sense of sadness at the results of the study. Dr. Cabot certainly must have hoped for better when he originated the program. It is upsetting from a humanistic point of view to learn that compassionate efforts to help may have such negative results. I trust that these conclusions do not turn people away from a concern with their society. The moral to be drawn is that we must work harder to make our knowledge commensurate with our compassion, and in the absence of knowledge, we must be duly humble in interfer-ing in the lives of others.

Early Warning

TheFourth of July seems to naturally evoke writing about the rights of man. We pick up a pen and ghosts of John Adams, Thomas Jefferson, and Tom Paine seem to be jiggling the other end urging us to declare that we have not given up on those principles that they so eloquently formulated. After feeling their urgings for many years, I have finally focused on the area in which I can personally most clearly see a present threat in the United States. The subject we shall consider is our long-cherished academic freedom, particularly with respect to scientific research activities.

Because things are often not what they seem to be, I must begin by recounting a statement a lawyer friend made to me in the early 1960s. He maintained that the frontline fighters in the battle for liberty at that time were the greediest of the businessmen, because they were always testing the limits of regulatory agencies. If unchecked by the avarice of these businessmen, the federal and state agencies would by fiat rob us of our independence, he argued. At that time I thought he was mostly joking, but I have become more and more impressed with the validity of his remarks as I have watched similar government-regulated erosions in the scholarly and research fields over the past 15 years. Except we academics lack the equivalent of greedy businessmen to protect us.

Visualize the private college or university as the veterans returned in 1945. The administration consisted of the president, the provost, and the deans, mostly professors who had been elevated into their administrative roles as a reward for successfully fulfilling their jobs. Decisions were directed at the needs of faculty and students, with a watchful eye on the purses of the alumni. The average university did no business with the government and jealously guarded itself from outside interference. Tenure even protected the radical professors from the wrath of the alumni. The nonfaculty managerial employes were often gentle refugees from the business world who saw no empires to build within the collegiate hierarchy.

Thirty years later a radical change has insidiously taken place. Within every university there exists an infrastructure, a large managerial cadre who exert appreciable influence on the operational affairs of the institution and are neither responsible to nor responsive to the faculty, the students, or the alumni. Who are these newcomers? How did they get their power? How are they eroding academic freedom?

They can best be identified by taking main listings from the directory of a representative American university. The listing begins with "Accounting, Associate Comptroller for," followed by "Administrative Data Systems, Office of," and continues through the alphabet to "Workers' Compensation"—having made no fewer than 49 stops along the way.

Among the Ps one would even find "Parking Department," "Power Plants," and "Purchasing Department, Division of Support Services." In addition to the university-wide administrative divisions, the major science departments and many other branches of the university have their own business managers.

Many of these offices were called into being either as a direct consequence of the university's receiving funds from the federal government or as a simple result of its being a major employer. An internal bureaucracy was evoked to interface with the corresponding federal structure. However, where funds are involved, power is not far behind, so that much of the decision making is shifting to the new managerial class. Academics and scientists, by the very nature of their dedication to abstractions, are only too glad to pass on all the annoying detail work to the helping hands. In the process of divesting themselves of specifics, they all too often fork over decision making that affects the very structure of the university.

Gradually, the nonacademic administrative arms expand the areas of involvement. Hiring must be done through Personnel. Buying of research and other supplies must be done by Purchasing. Remodeling of laboratories and classrooms must be arranged by Building and Grounds. Budgeting is largely done by nonacademic officials. The freedom of a principal investigator to operate is governed not by the demands of the research but by certain needs of the bureaucracy. The group leader must become a businessman entrepreneur and in so doing must inevitably yield to management decisions.

When compared to individuals in corresponding positions in the private sector, the scholar is inevitably disadvantaged. In the business world an independent activity generates capital that can be used to fight the government through lawyers, lobbyists, and bribes. Since the researcher is almost totally dependent on the government for support, he becomes a submissive tool in the hands of the system. The extreme example is the Soviet Union, where respected academicians regularly sign ridiculous petitions that everyone knows they don't believe. They sign, because to refuse is to remove oneself from one's chosen career. Things have not reached that stage in the United States, but which of us has not hesitated to sign a petition for fear of its eventual effect on our ability to get federal research funds? And which of us in the sciences has not geared the direction of his investigations to his perception of the granting agencies' requirements? Since research is the name of the game in universities, to be deprived of research funds is to be deprived of success. State institutions suffer all the problems of their private counterparts plus the addition of the state bureaucracy. At this point we should all hear the footsteps behind us.

It is difficult to sound so alarmist about a situation that seems so benign: scholars being directed and constrained by the managerial class. Yet an early warning system is imperative, what one of our novelists has called a "built-in, shock-proof crap detector." My personal radar is telling me, first, that academic freedom is a very fragile thing, and second, that managerial abuses have a constitutive, positive feedback that makes them constantly get worse in the absence of drastic countermeasures. Again, the Soviet example is so clear; it is written in the lives of distinguished researchers in prisons and insane asylums for daring to confront the system. In the United States we are being driven to "safe research" and "safe administrative procedures." Surely only a madman or a criminal would want to do things differently. And so we may fall into the bizarre situation that confronts Russian science. The early warning may not be early enough.

A freedom that is suddenly lost is missed and a cry of opposition arises. However, rights that are eroded day by day, not in the name of tyranny but in the name of such things as accounting procedures and fair labor practices, simply fade into obscurity and we awake one day to find that all is irrevocably lost. Liberty need not be surrendered to dictators; it can be nibbled away by bureaucrats, petty and grand, each supremely confident that he is protecting the integrity of the system.

The threat to academic freedom is not the mob from without, it is the white-collar mob within. They have been conjured up in response to taking the route of government funding. To refuse this route is to fall by the wayside; to accept it is to surrender independence of action. If it is not too late, the private universities as well as public institutions must somehow seek paths to preserve self-motivated action and integrity. To paraphrase Patrick Henry: Are federal grants so dear or success so sweet as to be purchased at the price of chains and slavery?

The Roots of Prejudice

In a more romantic age individuals were said to wear their hearts on their sleeves; now we seem to wear our minds on our automobile bumpers. It may not be a tribute to the depth of present-day thought that the messages come in such condensed form but, alas, we must take the world as we find it. Thus, after I had finally mastered the facts that "Sailors Have More Fun" and "Milk Drinkers Make Better Lovers," I was only mildly abashed by the injunction to "Protect Pure-bred Dogs, Buy from a Breeder." The fact that I was admonished to be concerned about the miscegenatious canine unions of every errant bitch in town came as a heavy burden, what with inflation and the energy crisis.

The whole question of breeds of domestic animals has been one of the fascinating chapters in the history of biology. All dogs are regarded as members of a single species, *Canis familiaris*, yet 116 different breeds are recognized by the American Kennel Club. A species is defined by the ability to interbreed, yet the thought of sexual relations between a Chihuahua and a Saint Bernard is perplexing to say the least, and might inspire wags. We can get into some serious conceptual problems in applying our biological definitions to domestic species.

The mechanism of variation under domestication occu-

172

pied even the great Charles Darwin. Chapter I of the celebrated *Origin of Species* dealt with this topic and formed a
cornerstone of Darwinian doctrine. The enormous variety
possible within a single species provided evidence of the biological variability which could lead to speciation. Dogs, however, proved to be an enigmatic case for the naturalist, who
noted, "I have, after a laborious collection of all known facts,
come to the conclusion that several wild species of *Canidae*
have been tamed, and their blood, in some cases mingled
together, flows in the veins of our domestic breeds." Pure-
bred, indeed. Other biologists, however, regard the dog as
a single line of descent probably from the wolf or the jackal.
In any case, dogs have lived with man for over 8,000 years
and have become the most variable of domestic animals.

Thus we have taken one or at most a small number of
varieties from nature and have bred them into 116 subspecies. Now, there are those among us who would guard the
purity of these subspecies to the point of an avid dislike of
all unsuspecting mutts who do not meet the Platonic ideals
ordained by the kennel clubs. Where will it all lead? What
of those families who share their bed, board, and love with
an individual animal whose genes do not come up to these
rigid standards? What if we applied the same principles to
humans, as is indeed done in some parts of the world where
caste and race are zealously guarded?

This radical turn of thought has no doubt been conditioned by events which occurred during a recent visit with
some old friends when we learned of prejudice against a
species of plant. To set the background, visualize a pleasant
small New Hampshire village which has seen many winters.
We are clearly in Robert Frost country, with a myriad of
birches to swing from, a host of unmended fences and abandoned mills along the river. Our hosts live in a postcolonial
frame house located in the center of town and surrounded
by gardens and a lawn. After we had reminisced about old
times for a while, the flow of conversation took us outside
to wander about the yard.

Our friend showed us the many things he had planted but reserved a special pride for his locust trees. He liked their rapid growth, their sturdiness, and the frilly quality of their appearance in winter. "It's a funny thing, though," he said, "most of the townspeople here don't like locusts. 'Weed trees' they call them in a kind of sneering way." And though not much else was said about it, in New Hampshire fashion, it was clear that this family had experienced some discomfort over their neighbors' reactions to the planting of locusts.

"Good grief," I thought, "species prejudice against the flora." Yet something about this particular breed of tree stuck in the back of my mind only to emerge when I returned home and noted in the encyclopedia that the black locust *Robinia pseudoacacia* is cultivated in Europe as an ornamental tree. There it was; the same plant that was scorned in New Hampshire was honored three thousand miles away in the parks and estates of "the old country." Truly the growth of the locust must tell us something about the roots of prejudice.

Doubtless all these ideas of trees and the dogs that urinate upon them would have passed quickly from memory had it not been for all the to-do over sociobiology, which has finally reached the mainstream of modern thought on the cover of *Time* magazine, although I have not yet seen it on bumper stickers. It has been a long and tortuous path from the Darwinian synthesis and the impact of humans facing their apeness to the modern controversy over humans facing their molecularity. I must confess I find the current argument something of a tempest in a teapot, for while it is certainly true we are the molecules that mold us, it is also clear that we are much more than these molecules. I feel completely confident in stating that organic chemistry will never explain why New Hampshire townspeople dislike a tree that is cultivated by their European relatives. A bit of moderation will surely show that while genes are most important in determining an individual's psyche and soma, there is no way that we can ever form a complete and reductionist view of man.

The physics upon which biochemistry depends is tied to the human mind. The biologist who wishes to reduce all to physics has not grasped the sense in which physics is a product of the biology of the mind. Some philosophical clarity would save a lot of argument.

However, thoughts of genes and behavior do open up some insights into the roots of prejudice. Let us take an illustrative case. When we hate or love an individual for something he has done to or for us, we are responding to ill-understood aspects of human behavior which characterize those phases of life that are most important to all. Facets of our feelings which we cannot get a rational grip on are truly the vital aspects of our lives. When, on the other hand, we hate an individual for his race, we are hating genes over which he has no control. The same is as true for humans as for mongrel dogs and locust trees. Prejudice means disliking the genes of a plant or animal or person without paying that much attention to the individual that goes with the genes.

Looked at in this light, prejudice gets stranger and stranger. When you get down to it, what we are hating are DNA molecules, nay not even DNA molecules but just the sequence in which the nucleotides are assembled. It seems odd that so much effort goes into venting our wrath on nucleotide sequences. It would appear that there are better things we could do with our time, what with inflation and the energy crisis.

Fair Is Foul, and Foul Is Fair

This is not an essay on ecology; rather, it is a commentary on scientists as people. It is, however, based on events that occurred at the First International Ecology Conference held in The Hague, Netherlands, in September of last year.

For the setting, visualize two thousand or so ecologists meeting in the Congress Hall at The Hague. It was truly an international meeting and one could identify participants from all areas of the globe. The fact that one man wearing a fez was from the Swedish delegation is a mystery I shall not attempt to penetrate. It was a serious meeting as international conferences go, for the business dealt not only with the abstract science of interacting species but with the pressing problems of how to protect the global ecosystem from the ravages of man and how to manage biological resources in the interests of man. Considerable thought went into water pollution and air pollution and their adverse effects on living organisms.

On Tuesday night the participants paused from their work to attend the prime social event of the conference: a reception in the Binnenhof sponsored by the Mayor of The Hague and the Netherlands Minister of Education, whose hospitality is deserving of praise. The site for the event was a large, truly palatial room, the Hall of Knights, with a central platform on which were mounted two thrones. Opposite the thrones was a long table manned by eight waiters who dispensed three kinds of wine, white, red, and rosé—all excellent. Waitresses circulated with trays laden with canapés.

The hall, which was built around 1300, was soon filled with invitees, and indeed the number of ecologists and "accompanying persons" soon crowded the room to a point of limiting mobility. My own "territory" was about four square feet of floor space, fortunately located next to the wine table. From this "nesting site" I was able to survey the "biome." The international character of the meeting was again most apparent in the clothing, the facial features, and the profusion of languages. The communication was catalyzed by the wine, and an air of camaraderie filled the hall.

And then, an event took place. From the east end of the hall, three men entered. They were dressed in black formal wear and each carried a tray with three objects: a box of cigars, a bowl of cigarettes, and a lighted candle. The symbolism struck me. From a medical point of view these men in black might just as well have been carrying bell, book, and candle. This trio of tobacconists then began to circulate among the ecologists and a strange thing happened. Wherever they went a cloud of smoke rose up until at last a miasma filled the entire room. The ecologists were eagerly grabbing at these cylinders of dried leaves of *Nicotiana tabacum*; they were then setting fire to one end of them and inhaling smoke from the other end while simultaneously causing clouds of smoke to billow into the surrounding air.

The net result was an effluvium that effectively penetrated every corner of the hall and made the habitat less than ideal for humans and other aerobic species. The fumes of the partially combusted leaves contained a rich variety of toxic organic compounds, the level of carbon monoxide was above threshold toxicity, and the particulate matter served as a mucous membrane irritant. This pollution of the local environment continued for the rest of the party. Finally, gin was served to signal the end of the festivities and we de-

parted the Knights Hall and found ourselves on the side-
walks.

I boarded a trolley to Scheveningen, the seaside town that
abuts The Hague. Standing on the boardwalk, inhaling the
air coming in off the North Sea, I tried to clear my lungs
and evaluate the meaning of the evening's happenings. The
group of men and women with whom I had spent the past
few hours were accomplished professionals, of obvious intelligence and success. They were, of all the people in the
world, among those most concerned with environmental
quality. They spent their days debating how to manage the
world so as to protect it from its excesses. Yet their behavior
at the Knights Hall that evening could best be described by
Macbeth's witches who proclaimed:

> *Fair is foul, and foul is fair:*
> *Hover through the fog and filthy air*

The conclusion, which is a recurring theme in human affairs, is that there is a large gap between the intellectual
realization of what must be done and the will to carry through.
Every person in the Knights Hall knew that tobacco smoke
pollutes the air and in a closed room reduces the quality of
the environment. Yet almost no one responded to that
knowledge by the obvious act of simply passing up the tray
of cigars and cigarettes and settling down to a very pleasant
evening. The solution is in the Biblical proverb "Physician
heal thyself," which we might paraphrase "Environmentalist
purify thyself."

In "Ligeia," which Edgar Allan Poe called "my best tale,"
the beautiful Ligeia dies murmuring the words of Joseph
Glanvill: "Man doth not yield him to the angels, nor unto
death utterly, save only through the weakness of his feeble
will." So mankind will not yield itself to crisis and despair
save if its collective will is too feeble to survive. The collective
will of the Knights Hall ecologists indicates that we have a
long way to go.

ered they kingdom and finished it," we must also be alert
o being numbered by overpopulation, overindulgence, and
concomitant shortages.

Having arrived in the Bay area from the East Coast where
water was plentiful (indeed, we were having floods), we found
the impact of insufficient water particularly dramatic. It may
be belaboring the obvious to note that shortages are remote
abstractions until experienced. Thus while reams are written
about the energy shortage, for example, it is only taken seriously when factories close or long lines form at the gasoline
station. So it was with knowledge of the drought. While the
newspapers proclaimed its existence and further discoursed
on the threat to agriculture, the fact that toilets weren't being
flushed was never mentioned. People in wetter parts of the
country were also unaware that Californians were: saving
bathwater to put on the garden, changing their style of show-
ering (no snickers please), and undergoing what would be
called a downgrading of living standards. No one was suffering or severely hurt, since our absolute biological requirement is about one and a half liters per day, and in the worst
areas 160 liters per day per person were available. Nevertheless an entire population was experiencing a change of
life-style. I suppose that the deeper message of the bathroom
graffiti is a reminder that shortages, present and future, had
better be envisioned in earthy terms if we are going to respond effectively.

Since the beginning of population centers, water supply
has been almost synonymous with civilization. Thus, besieg-
ing cities in biblical times often meant controlling the wells.
The golden age of Greece was accompanied by a technolog-
ically advanced communal water supply. The aqueducts of
the Romans still stand as an engineering masterpiece of antiquity. Starting with the Aqua Appia in 312 B.C., the system
eventually grew to 359 miles total length in 600 years. The
fall of Rome led to decay in water supply, and the Dark Ages
were unenlightened with respect to providing the public with
clean, pure H_2O. The modern high-pressure systems have

Yellow Is ▊

One supposes that thoughts may arise at any time ▊ or night and those that come to the fore during ▊ physiological activity need be no less noble than the one▊ come to mind in more esthetic surroundings. Perhap▊ special role of graffiti as a literary form flows from the ▊ that, philosophy aside, the bathroom is truly a point of ▊ tact with reality, and lesser distinctions are blurred in ac▊ ities common to all. Suffice it to say that we ▊ thermodynamically open systems, a fact of existence whi▊ is inescapable.

Thus I was brought to serious contemplation of the Wes▊ Coast water shortage when I read on the wall over a Berke-ley, California, urinal the message, "If it's yellow it's mellow, if it's brown, flush it down." If this handwriting on the wall was more whimsical than the classical *mene, mene, tekel, uphar-sin* on the wall of Belshazzar in the Book of Daniel, it was equally ominous. For when water became scarce, one of the first things to go was the luxury of emptying the bowl after each usage. Thus the great invention of Thomas Crapper (see *Flushed With Pride* by Wallace Reyburn, Prentice-Hall, 1971), the flush toilet—one of the truly substantive advances in public health and public esthetics—was falling victim to the drought. If Daniel interpreted *"mene"* as "God hath num-

been developed since the late 1500s and purification procedures began only about 200 years ago. A full rational theory of water supply required a knowledge of bacteriology,
and the necessary trace components are still a matter of dispute.

Trying to focus on our need for water leads to a consideration of that substance in the life and very material makeup
of man. Western philosophy is supposed to have begun when
the Ionian philosopher Thales proclaimed the unitarian
principle that "all things are water." Into the modern period
one of the great odes to the special properties of H_2O in
living systems is found in that perceptive book, *The Fitness of
the Environment*, written in 1913 by Harvard Professor Law-
rence J. Henderson. He deals extensively with water and the
physical properties that make it unique as a biological medium. For example, there is the very high specific heat which
stabilizes the temperature fluctuations of oceans, lakes, and
streams and moderates the changes that organisms are subject to. Most living things are 50% to 95% water, and temperature homeostasis is favored by a large heat capacity.

Other thermal properties noted by Henderson include the
high latent heats of the solid-liquid and liquid-vapor transformations as one goes from ice to steam. These energy sinks
stabilize environmental temperatures and allow evaporation
to serve as a refrigeration mechanism to rid organisms of
excess heat produced in metabolic activity. In dogs at 35°C,
a full 79% of the heat loss is accounted for by the latent heat
of vaporization.

Ordinary aqua pura has a maximum density at 4°C above
the freezing point, as well as a large volume change on freezing. Thus ice and the coldest liquid float on top of bodies of
water. If freezing took place from the bottom, Henderson
reasoned, the ice would form at the lowest levels each winter
until it gradually built up, making the sea or lake too cold
for habitation.

In addition to its strange thermal properties, H_2O has a
very high surface tension and dielectric constant. The first

of these accounts for the rise of fluids in capillary systems, which explains certain features of soil as well as the rise of sap in tall trees. The dielectric constant determines the solvent properties and is important to cellular function. High dielectric constant results from the pronounced electrical asymmetry of the water molecules, with the oxygen carrying a high negative charge and the hydrogens a high positive charge. The great electrical difference between water and hydrocarbons is behind the general notion that oil and water don't mix, and eventually is involved in the formation of lipid bilayer membranes in cells and organelles.

Henderson summed up with the thought: "Water, of its very nature, as it occurs automatically in the process of cosmic evolution, is fit, with a fitness no less marvelous and varied than that fitness of the organism which has been won by the process of adaptation in the course of organic evolution."

The theme outlined above was taken up again in 1958 by Harvard Professor John Edsall and his coauthor Jeffries Wyman in their book, *Biophysical Chemistry*. (Harvard folks seem to show a healthy interest in water.) They go beyond Henderson in discussing the molecular structure of water, which was not known at the time of the earlier book. Of particular interest is the hydrogen bond which is responsible for most of the anomalous properties of the liquid phase. They also conclude "the structure of water is truly unique."

From Thales to the present day, the remarkable water substance has been at the center of biological thought. That writing about this material has moved from distinguished monographs to graffiti only indicates that concern over the supply of water is much too important to be left to the scholars and now is everyone's business. And it had better be everyone's concern before the whole civilization goes down the drain.

Extraordinary Popular Delusions

I am one of those Americans who has not seen the Tutankhamun exhibit. I make this statement without a sense of pride or remorse. A lack of enthusiasm for crowds and the idea of standing in lines months in advance of the exhibit were enough to decide the matter. I do, however, recall with pleasure several interesting times spent in the Egyptology sections of the Metropolitan Museum of Art, the British Museum, and other institutions. All of these displays were relatively uncrowded except for a cocktail party among the sarcophagi at the University of Pennsylvania Archeological Museum in Philadelphia. What has always impressed me about early Egyptian culture is the obsession with death displayed in the massive amounts of funereal art and architecture. The creative thrust of that society was "interred along with its bones."

Obsession with death is much in the air lately since the miasma from Jonestown, Guyana, has spread its way around the world. But the Tutankhamun exhibit and Jonestown have more in common than death itself; they are in very, very different ways examples of crowd madness, our present focus of attention.

In 1841 an English author named Charles Mackay wrote an extraordinary book entitled *Memoirs of Extraordinary Pop-*

ular Delusions and the Madness of Crowds. The 1852 edition begins with the lines:

> In reading the history of nations, we find that, like individuals, they have their whims and their peculiarities; their seasons of excitement and recklessness, when they care not what they do. We find that whole communities suddenly fix their minds upon one object, and go mad in its pursuit; that millions of people become simultaneously impressed with one delusion, and run after it, till their attention is caught by some new folly more captivating than the first. We see one nation suddenly seized, from its highest to its lowest members, with a fierce desire of military glory, another is suddenly becoming crazed upon a religious scruple; and neither of them recovering its senses until it has shed rivers of blood and sowed a harvest of groans and tears, to be reaped by its posterity.

And so Mackay chronicles: great financial follies such as the Mississippi Scheme in France and tulipomania in Holland, religious activities such as the Crusades and witch mania, various esoteria such as fortune-telling and haunted houses, and strange behavior such as the slow poisoners and magnetisers. Each of these has led to excesses which have swept up and destroyed individuals and nations.

Tulipomania in Holland is one of the most instructive of Mackay's examples. In the early 1600s, tulips became so prized that vast sums were paid for a single bulb. Tulip markets arose in the larger cities and speculation in these flowers became a major financial activity of the country. The fad came about so rapidly that returning travelers often did not know of it. A seaman, home from a long voyage, noticed a bulb, looking very like an onion, sitting on the counter of a merchant. He walked off with it, not realizing he had a rare variety of tulip. He was found munching on the specimen and a herring, unaware that his breakfast was valued at 3,000 Florins. This resulted in many months in prison on a grand larceny charge. All madness eventually comes to an end; the tulip market crashed, ruining a large number of speculators.

The sadder, and one hopes wiser, Dutch spent the next several years putting their economy back in order.

Since first reading Mackay's book many years ago, I have regarded it as one of the most precious volumes in my library. For in informing us of the phenomena of crowd madness Mackay prepares us to resist the urges of the herd and to live saner lives truer to our inner selves. Commitment to this concept was demonstrated a number of years ago when my wife and I bought copies of this book to give to each of our children so that their libraries would also contain this antidote to popular delusions.

Tutomania as distinguished from tulipomania is the movement of millions of Americans into museums. I doubt that they have come because of an abiding interest in Egyptology; they were not there crowding the classical civilization sections of the museum before this exhibit and few will be found there after the show closes. They have materialized as if by magic, evoked by the electromagnetic waves of television and pumped into existence by the rivers of printers' ink. It is a harmless mania; we might go further and classify it as a beneficial movement. Therefore it can be studied without rancor as an example of crowd madness *per se*, fortuitously there to show us people responding to a popular craze rather than to something of intrinsic interest. We should study it carefully, think about it and tuck the knowledge away, so that when a malevolent crowd madness occurs we will understand and be able to deal with the kinetics of such a demonstration.

Jonestown was an example of pernicious crowd behavior. The individual members of the People's Temple lost their sense of self even at the biological level of survival. As a result it is almost impossible to imagine what went on in that jungle community.

Recent history has presented this century with an example of crowd madness that would doubtless have shaken even the skeptical Dr. Mackay. We have witnessed one of the great civilized nations, the intellectual heirs of Bach, Goethe, Kant, and Helmholtz, caught up in a frenzy which drenched the continent of Europe in a sea of blood that sweeps over our ability to comprehend the events. Thus the tales related by Mackay are but a sample of this genre of insanity. The author described his work as "a chapter only in the great and awful book of human folly which yet remains to be written and which Porson once jestingly said he would write in five hundred volumes."

I must confess that I have on numerous occasions put down the daily newspaper and browsed through my copy of *Extraordinary Popular Delusions*. It is not as worn a volume as the Bible or my *Introduction to Theoretical Physics*, but nevertheless it is beginning to show the effects of many rereadings. The last heavy use occurred during the swine flu vaccination program. In my most cynical moments (which are fortunately rare) I envisioned thousands of physicians injecting millions of individuals with an untested cure to a nonexistent disease. I went back to Mackay's section on the alchymists and thought maybe I should send a copy to the Center for Disease Control so that they could check out whether things were "All strange and geason/Devoid of sense and ordinary reason."

Crowd madness can strike any nation, any profession, or any subgroup in a society. The first step in dealing with the phenomenon is to recognize it, and if I have been a bit hard on the Tutomaniacs it is just to alert you to the severer forms of this type of behavior, concerning which we must be eternally vigilant.

Frankenstein and Recombinant DNA

Having been thoroughly frightened as a youngster by Boris Karloff's cinematic rendition of the monster, I was hardly prepared for the contents of *Frankenstein or the Modern Prometheus* written in 1816 by Mary Wollstonecraft Shelley. The book was strongly recommended to me by a student as a volume of combined literary, biological, and philosophical interest. That theme was reinforced in the first line of the preface where Erasmus Darwin, biologist and eccentric grandfather of Charles, was referred to as supporting the possibility of the occurrences of the events portrayed.

The genesis of the novel is as fascinating as the contents. A group of four friends on vacation near Geneva huddle around a fire telling ghost stories. They each agree to write such a tale. This remarkable ensemble included Lord Byron, Percy Bysshe Shelley, and Mary herself, a young lady of only nineteen years. The fourth member was the enigmatic John Polidori, Byron's young physician and traveling companion. Starting from this whimsical challenge, Mary Shelley created a myth which has been the source of theatrical inspiration through the years. She also crafted a moral argument which reechoes today for those willing to hear the sad and plaintive voice of Victor Frankenstein.

To set the background for the ethical confrontation we

must retell part of Mary Shelley's story, starting with the arrival of the young student at the University of Ingolstadt. He was immediately drawn to the sciences and commenced an intense two-year program which culminated in the study of anatomy and an obsession with the processes of death and decay. Then in a flash of blinding insight he discovered the secret of how to animate inanimate matter. [In the interest of our philosophical discussion, we must ignore the bad science set forth by the author. Note that the book was written well over a decade before the cell theory.] He proceeded to reduce his idea to practice, noting "The dissecting room and the slaughterhouse furnished many of my materials." His obsession became a mania, and he worked for a year, day and night, to produce an eight-foot-tall living being. The large size was necessitated by the crudeness of miniaturization techniques. At last he succeeded and "saw the dull yellow eye of the creature open."

But his success was only partial. In his fierce concentration on the sciences he had ignored the arts, and the creature he had fashioned was, when animated, so incredibly ugly and horrible it became "a thing such as even Dante could not have conceived." Frankenstein abandoned his creation, fled the laboratory, and wandered the town. When he returned the semihuman giant was gone, and Frankenstein, falling ill, simply abandoned the enterprise with no thoughts as to the future of his "child." Throughout the rest of the story until its ultimate climax the scientist never realized that it was not the biosynthesis alone that caused the tragedy, but the synthesis followed by abandonment.

And so a year passed with Victor shutting out of his mind thoughts of what had happened. His retreat from reality ends when he learns that his younger brother has been murdered, and he returns home to Geneva to find that the instrument of the boy's death was the very being fashioned by his own hands. Shortly thereafter Frankenstein and the monster meet on the lower slopes of Mt. Blanc. A strange monologue then ensues in which the humanoid narrates his short life story. The central theme of this biography is a desire on the part of the monster for human interaction and friendship. This desire is constantly thwarted by rejection on the part of people who respond to the ugliness of the stranger in cruel and often violent ways. Thus the naiveté of the newly formed being gradually turns to hatred and despair. The culmination was the unplanned murder of Victor's younger brother.

At this point the narration ends, and the creature offers to make a deal with the scientist. If Frankenstein will synthesize a female of the "same species and defects" to provide a companion for the monster, the two humanoids will go off to some unpopulated part of the world to live out their lives in peace. If the scientist refuses, further revenge and destruction will be wreaked on him and his loved ones. Frankenstein agrees to make a female, and subsequent events lead him to a remote Orkney island off the coast of Scotland where he has set up a laboratory in a small hut. While working at yet a second creation of life, the scientist ponders the consequences of his acts and comes to the horrible realization that

> Even if they were to leave Europe, and inhabit the deserts of the new world, yet one of the first results of those sympathies for which the daemon thirsted would be children, and a race of devils would be propagated upon the earth, who might make the very existence of the species of man a condition precarious and full of terror. Had I a right, for my own benefit, to inflict this curse upon everlasting generations? I had before been moved by the sophisms of the being I had created; I had been struck senseless by his fiendish threats: but now, for the first time, the wickedness of my promise burst upon me; I shuddered to think that future ages might curse me as their pest, whose selfishness had not hesitated to buy its own peace at the price perhaps of the existence of the whole human race.

Frankenstein decides he cannot go ahead. He destroys the

half-finished female, demolishes his laboratory, and sets forth on a terrifying personal tragedy which culminates in his own death and the self-destruction of his miscreation.

Mary Shelley's words make one recall, with a Gothic terror all their own, the grim warning set forth 150 years later by one of the modern pioneers in elucidating the structure of DNA. Erwin Chargaff wrote (*Science* 192:940, 1976):

> Have we the right to counteract, irreversibly, the evolutionary wisdom of millions of years, in order to satisfy the ambition and curiosity of a few scientists?
>
> This world is given to us on loan. We come and we go; and after a time we leave earth and air and water to others who come after us. My generation, or perhaps the one preceding mine, has been the first to engage, under the leadership of the exact sciences, in a destructive colonial warfare against nature. The future will curse us for it.

How strangely similar are the moral arguments set forth by a 19-year-old novelist and a mature scientist a century and a half later. Yet how different are the responses. Frankenstein sacrificed his personal goals to the social good. Contemporary researchers are criticized by Chargaff for rationalizing their own aims and deciding that they are indeed social goals, good for all of us. The earlier savant realized that he was being moved by the sophisms of the structure he had created. We are swept up in our own paradigms and their sophism and thus decide that what is good for the curiosity of the scientific community is good for the world. It is perhaps time for scientists to step back from their own arrogance and take another look at themselves. Are we indeed exercising the moral depth of a young ghost-story writer, let alone the far greater depth demanded by the magnitude of the question?

I have for several years now been haunted by the moral imperative to make a statement about recombinant DNA. The long delay has been caused by the failure to find the language to express this overwhelming issue adequately. The words of Victor Frankenstein and the horror of Mary Shelley's story seem to capture the essence of the argument better than any sermonizing that we might do. The possibility that our mistakes might self-replicate adds new dimensions to the care with which we must judge the products of our labors. I fear that most investigators are insensitive to these problems, acting more like the sorcerer's apprentice than the tortured unhappy soul wrestling with his conscience on a lonely Scottish island.*

*Since writing this essay, I have found reference to a rabbinic legend, much more ancient than the Frankenstein story, which reports, "The leviathan was created on the fifth day. Originally God produced a male and a female leviathan, but lest in multiplying the species should destroy the world, he slew the female, reserving her flesh for the banquet that will be given to the righteous on the advent of the Messiah." (*The Jewish Encyclopedia, Funk and Wagnalls Co. 1916*)

PART VI

The Wine of Life—
PORING OVER IDEAS

The Wine of Life

On the evening of February 10, 1978, a group of distinguished scientists gathered at the Stanford University
School of Medicine to commemorate the 100th anniversary
of Claude Bernard's death. Two days of meetings followed,
exploring the current status of fields in which the French
physiologist's discoveries played a leading role in developing
the modern point of view. Such occasions lead us to pause
and consider the individual being remembered. The theme
for such thoughts was set by Plutarch some 1,900 years ago
when, writing of Alexander the Great, he noted, "The most
glorious exploits do not always furnish us with the clearest
discoveries of virtue or vice in men; sometimes a matter of
less moment, an expression or a jest, informs us better of
their characters and inclinations."

Physiology and medicine was not the first career choice of
Claude Bernard; rather, he entered this field following his
failure as a playwright. In 1834 he submitted his play, *Arthur
of Brittany*, to be read by Sorbonne Professor Saint-Marc Girardin. The critic noted that the young man lacked the "tem-
perament of a dramatist" and suggested that he study
medicine. Each life seems to have choice points where unexpected events influence the path. Such happenings are
unpredictable and should induce a sense of humility in those

charged with advising the young. In the case of Bernard, Girardin's advice was physiology's gain, and we have no way of knowing what kind of dramatist the young man might have matured into. Later, as a philosopher of science, Bernard was strong on experimental controls, but life offers no controls for "the road not taken."

Although Bernard's scientific life was an unending series of successes, his family life seems to have been particularly empty. At age 32 he entered into an arranged marriage with Marie Francoise Martin. The bride brought a dowry of 60,000 francs, and this sum allowed the researcher the opportunity to remain in Paris rather than return to his home village to practice medicine. The dowry served in a sense as a research fellowship, in our terminology, but a high price was paid for the grant. The marriage produced two daughters who, like their mother, became alienated from Bernard. Perhaps the most poignant aspect of this alienation was Mme Bernard's activities in supporting antivivisectionist causes. While he was solidifying the science of physiology with animal experiments, she was advocating the banning of this type of work. Such divergence was hardly the foundation of a lasting marriage. The union was finally terminated by a permanent separation.

When, after a vigorously active period of years in the laboratory, illness forced Bernard to rest from the rigors of experimental research, he retired to the country and wrote his best-known book, *An Introduction to the Study of Experimental Medicine*. Though hardly indicated by the title, the work became recognized as a study in the philosophy of science, and no less a philosopher than Henri Bergson regarded it as a significant contribution. It was used as a text in philosophy courses in France, and some have regarded it as comparable in significance to *Discourse on Method* by René Descartes.

I think it interesting to note the number of scientists who, after a successful career in research, turn their attention to philosophical issues. This has perhaps characterized physicists more than biologists—think of Einstein, Schrödinger, Wigner, and Elsasser. But the example of Bernard, as well as others like Julian Huxley and Edmund Sinnott, shows that this phenomenon is general.

Science and philosophy have had a curious love-hate affair for the last 25 centuries. The word metaphysics itself was introduced to describe that material that came after the physics in the works of Aristotle. Metaphysics has been both condemned as a barrier to science and sought after to properly establish the foundations. It seems most unfashionable when a given discipline is on a wave of success and most popular when paradoxes arise to challenge a set of established procedures. The turbulent state of this relationship will doubtless continue for many arguments to come.

Somehow, in the fire of youth the sharp challenges of science seize the mind of researchers and drive them to solve defined problems. After a number of successes in this arena scientists often seem to ask themselves: What does it all mean? How do my hard-won results fit into a broader pattern? Do we understand man and the universe better now than when I began my researches? For some, such as Bernard, this becomes a time of reflection, either to set forth views on method and analyze the research process or to probe the edges and explore more speculative areas. The urge to philosophize becomes dominant in many as they begin to muse on their own mortality.

Moving, however, from mortality to nativity, we note that Claude Bernard was born in 1813 in the village of St. Julien in the Beaujolais region of France. The very name of his birthplace is apt to tickle the palates of wine lovers, and indeed throughout Bernard's entire life he had a profound interest in the growing of grapes and their fermentation. His father was a vine grower, and that tradition went back in his family for many generations. After he moved to Paris he came back to his home every year for the vintage, the gathering of the grapes and the preliminary process of wine making. When Bernard was an older man, war once kept him from returning, and he wrote, "This is the first time in my life that anything has happened to prevent my being on hand for the vintage in my own countryside." It is perhaps no coincidence that some of his earliest researches dealt with the digestion of sucrose.

In 1860 he bought a house and vineyard near the family home. He regarded the vineyard as an investment in his financial future, and he like his ancestors became a grape farmer. In 1869 he wrote to a friend that for a time he had become transformed into a wine maker. He went on, "These are the occupations which are familiar to me, and in the midst of which I was born; they always give me pleasure and they are certainly much more agreeable to me than composing academic discourses." Along with his other accomplishments Bernard was an enologist in both a romantic and a technical sense.

His notebooks show that as early as 1851 the wine lover was beginning to think about the scientific aspects of fermentation. By 1873 he undertook some experiments on alcohol production. The subject at that time was dominated by Bernard's younger contemporary, Louis Pasteur, who believed that the process depended on the presence of living yeasts and the absence of oxygen. Bernard's experiments appear to have been directed toward attempting to isolate a soluble enzyme system free of living yeast cells. This approach followed quite naturally from the success he had experienced with soluble digestive enzymes. The last experiments carried out by the noted physiologist were on the fermentation of grape juice. He remarked that Pasteur has seen only one side of the question.

After Bernard's death in 1878 some fragments of his observations were published and these produced a posthumous dispute between the two preeminent 19th-century French biochemists. Pasteur had the advantage of being alive and able to refute Bernard's notes. Bernard had the advantage of being correct in his guess that enzymes in solution could cause alcoholic fermentation, although his experiments certainly did not prove his case. The matter was finally resolved in 1895 when Eduard Buchner prepared the soluble enzyme system.

The dispute is now a matter of the past. What stands is the whimsical idea that the ultimate confrontation between these two giants of French science revolved around the issue of how alcohol is produced. Wine making has now spread over the world, and French wines no longer totally dominate, but no other country will experience a disputation such as that between the ideas of Louis Pasteur and Claude Bernard on how grape juice becomes wine.

Splitters and Lumpers

The terms "splitters" and "lumpers" come from taxonomy, where the classifiers were separated into those who liked to create new taxa because of small differences and those who preferred to coalesce categories because of similarities. The concept has found wider applicability as knowledge in all fields expands. Specialists are confined to ever-narrowing domains while generalists survey the immensity of information in an effort, one hopes, to find higher orders of structure. It is clear that in the university and intellectual community, entering the last quarter of the 20th century, the splitters are in command and the lumpers are in serious disarray, unable to keep up with the output of printouts that are generated in such a wide variety of ways. It is saddening to witness the loss of status of those engaged in integrative thought, for one sees in it the fragmentation of scientific and humanistic disciplines. The importance of the problem was set forth by the late Erwin Schrödinger, himself a scholar of great range. He noted that "it seems plain and self-evident, yet it needs to be said: the isolated knowledge obtained by a group of specialists in a narrow field has in itself no value whatsoever, but only in its synthesis with all the rest of knowledge and only inasmuch as it really contributes in this synthesis something toward answering the demand 'who are we?' "

Both the splitters and the lumpers are essential to the development of any intellectual edifice. The acquisition of new facts is usually the work of the former, while the organization of material in a form communicable to a wider group of thinkers is the task of the latter. The question before us is not the validity of either approach but the proper balance of efforts in the continuing quest for understanding. It is also important to judge both groups by the best they are capable of. Splitters' syndrome, which is a propensity for meaningless busywork, is not the mark of the great particularists. Neither should comprehensive scholars be judged by those refugees from detail who, having failed in their own efforts, hide behind a facade of sophisticated-sounding nonsense in order to mask the emptiness of their activities.

The problems of formulating a global view cannot be underestimated. There are those who maintain that we must look to Helmholtz or Spencer to find someone with a comprehension of the learning of his day. Others would claim that we must go back to Leibnitz to find such an individual. It is doubtful whether we shall ever be able to return to the ideal of Renaissance Man—one versed in all the arts and sciences. How then can the intellectual community contribute to the restoration of a sense of wholeness? The first step is to believe in some kind of unity underlying the vast phenomenological diversity of the modern world. The second step is to provide a setting for modern-day thinkers to pursue more holistic approaches to the enormous output in the specialized literature.

The situation we focus on is the absence of sufficient integrative effort within the structure of the modern university. The sociological causes are many, but the immediate causes are fourfold: the difficulty of evaluating the competence of generalists, the crowd of charlatans who have given synthesis a bad name, the absence of niches for wide-ranging thinkers in the departmental structure, and the flow of funding toward the concrete.

When an individual is doing a given task that is directed

and well defined, it is a quite simple matter to evaluate competence and productivity. For a synthesizer, however, the results of the effort may only be available after a long time; maturing may be a slow process, and a sharp eye is necessary to distinguish valuable efforts from impressive-sounding and well-merchandised nonsense. The universities have as a rule responded to this problem by taking the safe route of filling chairs with accomplished specialists who are recognized with ease by a "society" of similar individuals. It is a simple job to ask for competence; wisdom is a far more difficult matter.

How one judges the success of a lumper is something upon which lumpers themselves cannot agree. Witness the following two statements on Teilhard de Chardin, each made by an established generalist in science:

Jacques Monod—*Although Teilhard's logic is hazy and his style laborious, some of those who do not entirely accept his ideology nevertheless allow it a certain poetic grandeur. For my part I am most of all struck by the intellectual spinelessness of this philosophy. In it I see more than anything else a systematic truckling, a willingness to conciliate at any price, to come to any compromise.*

Julian Huxley—*His influence on the world's thinking is bound to be important. Through his combination of wide scientific knowledge with deep religious feeling and a rigorous sense of values, he has forced theologians to view their ideas in the new perspective of evolution, and scientists to see the spiritual implications of their knowledge. He has both clarified and unified our vision of reality.*

Given such radical difference among the experts on assessment, it is small wonder that only the boldest and toughest of university leaders would attempt to evaluate universalists.

Note, however, that in the end, when we abandon the search for catholic thought in our institutions of higher learning, we see it reappearing elsewhere in a series of movements such as Arica, Silva mind-control, Scientology, and Erhard Sensitivity Training. The yearning for wholeness does not go away, but when ignored by disciplined scholars it takes on bizarre forms in the minds of those far less concerned with intellectual and logical rigor. The negative act of ignoring integrative thought in the universities is the positive act of leaving the field free to those whose commitment to knowledge may be secondary to other goals.

I must confess that my own undergraduate years were so enriched by two of the world's great generalists—Henry Margenau (*The Nature of Physical Reality*) and F. S. C. Northrop (*The Meeting of East and West*)—that I have a strong bias toward the value of such thinkers in the educational process. However, I have also known a sufficient number of the other kind of so-called generalist to have a firm sense of the suspicions harbored by those laboring away at more specific tasks. It is painful to see two sides of a discussion so clearly.

At this point we are in the embarrassing position of having raised an issue whose solution does not emerge. The evaluation of generalists is, perhaps, a problem on which a bit of progress is possible. We can demand that an individual achieve a certain success in a particular discipline as a prerequisite to accepting him as a holist. It is the equivalent of demanding that an individual be a successful draftsman before we allow him to be a nonobjective artist. Synthetic thought under these conditions may become the activity of tenure professors who could share their insights with undergraduates who have still not entered the trade-school phase of their careers. All in all, that might not be a bad start; it would illuminate the choice of specialization for the young and allow maturer scholars the chance to think about more global questions. The intervening years could be concerned with the work of the world. What we then ask of the universities is to spare the wide-ranging thinkers a certain derision that they may at present experience from those engaged in more directed tasks. We need to reassert that the job of the university is to search for the nature of man—who he is and where he is going. It is a serious business, and we must struggle to avoid what the philosopher Ortega y Gasset has called *la barbarie del'especialismo*.

Bibliophilia

The stacks of any major library are a unique environment. Dimly lit and architectonically partitioned to achieve the maximum ratio of shelf area to volume, they reflect an eeriness appropriate to such repositories of knowledge. The users pad by in hushed silence and gnomelike peer myopically at the call letters. It is not a place for casual conversation, and I cannot recall ever having formed an acquaintanceship in such surroundings.

Yet I must confess a certain fondness for these darkened vaults. They are the traces of human culture, the storehouses of past wisdom and folly. And there is always the chance, ever so slim, that one will pull a volume down off the shelf, open the cover, and *voilà*—enlightenment. There are perhaps deeper reasons for this bibliophilia, going back to my youth. On many occasions when my peers were off playing ball or going to the movies I would wander through the stacks of the hometown library looking for that certain illuminating volume. On one such afternoon I busied myself estimating the number of books in the stacks and came to the sober conclusion that in a lifetime it would not be possible to read them all. That was a loss of innocence, and intellectual life has never seemed quite the same since.

Returning to the present, vacation had come and there I was in the bowels of the medical school library looking for some "quaint and curious volume of forgotten lore." The sub-sub-basement was dirty, and the layers of dust on the books were thicker than I had remembered them. Many volumes were out of place, and I had the greatest difficulty in trying to locate the desired journal articles. I recalled a similar experience a few months ago during a visit to another university. In the stacks of that main library there was an overabundance of dirt and disorder. In both cases the reasons were clear; insufficient funds and mismanagement left too few to counter the universal entropic drive, and the inexorable processes of nature were doing their work.

However, a disordered library is a useless heap of books. A misplaced volume among several million is effectively lost. This kind of scholarly resource has no permanence and, like all other products of human effort, must be renewed and maintained lest it dissipate and fall to dust. So it was with a sense of gloom that I surveyed the surroundings and sat down to read one of the works that I had at last been able to find.

Somehow it was very difficult to concentrate on the cytological matters on the reading desk before me as the mind wandered in space and time to the Library of Alexandria, one of the unsung wonders of the ancient world. The fate of that institution as well as the other great collections of antiquity kept intruding itself upon me. The message was obscured, but the obscurity was itself part of the message. The reason we know so little about the libraries of the ancient world is that their destruction has robbed us of that knowledge.

We do know that Alexander the Great, having designated the site of his western capital, then moved on in his unending pursuit of empire. He died without ever having seen the royal city that rose on the shores of the Mediterranean, and shortly after his demise one of the Macedonian generals arrived to take over the government of Egypt. He became Ptolemy I, and established in Alexandria a dynasty that persisted until the death of Cleopatra.

Alexander's military aides, who were regarded as crude barbarians by the effete elite of Athens, were nevertheless thoroughgoing Hellenists whose zeal for conquest was almost matched by a desire to spread the culture of Greece. Thus after Ptolemy I solidified control he established the Museum (University) and the Library. The great classical manuscripts were then purchased, purloined, or pillaged, and gathered in a large marble building. For the next three hundred years all of the successive Ptolemies added to the collection while the scholars of the Museum edited, redacted, translated, transcribed, and contributed to the classical works. When Julius Caesar arrived in 48 B.C. the Library boasted 700,000 scrolls.

But as I was sitting in the dirty disordered stacks my thoughts could not long remain on the glorious rise of the Library of Alexandria, drifting rather to the lamentable decay of that institution. The first assault on the collection came when Cleopatra presented to Caesar as many as 40,000 manuscripts. We are uncertain as to their fate, but there are some reports that they burned in the dockside warehouses when Roman legions set fire to the Egyptian fleet. Cleopatra's couch was not only the check-out desk but on occasion the acquisitions department of the Library, as when Marc Antony stole 200,000 scrolls from Pergamum and presented them to his love. Sic transit gloria Pergami.

Through imperial depredation and civil wars, the Library appears to have survived intact from Cleopatra's death for another three hundred years, although biological processes such as rot, decay, worms, and insects were taking their toll. These natural forces were then abetted by war and invasion. The Museum was destroyed, and the Roman masters ordered that all the ancient books which dealt with the art of processing gold and silver be committed to the flames. This selective book burning was an attempt of the conquerors to control the Alexandrians by depriving them of the technology of coinage.

Almost a century later the Library suffered its next major

trauma. Christianity triumphed and an edict went forth from Rome for the destruction of paganism. The prelate of Alexandria was a violent man, one of the bishops that the Church would just as soon forget about. He led a mob which reduced to rubble the contents of the Temple of the God Serapis. Along with the Serapeum went many scrolls, part of the collection of the Alexandrian Library. Mankind's search for God had taken its toll, a theme whose variation was to be replayed in 300 years.

The final chapter in our historical drama occurred in 640 when an Arab army of devout Muslims conquered Egypt. The commander wrote to Caliph Omar asking what to do with the Library and received the following reply: "Regarding the books of which you have told us, if they contain anything which conforms to the Koran, the Book of God permits us to eliminate them; if they contain anything which is contrary they are useless; proceed then to destroy them." Legend has it that they were burned to heat the public bath of Alexandria. Once again information was reduced to heat, the ultimate form of energy. We are thus reminded that most of human knowledge lies somewhere between redundancy and heresy.

While thinking about all this was somewhat depressing, it did lead to a much deeper appreciation of the surrounding materials. Before leaving my library, I tried to straighten up the bound volumes of one journal, and on the way out I picked up some trash from the floor and put it in a wastebasket. Not much, but at least a move in the right direction. Libraries perish for many reasons. They persist only if a society makes a firm commitment to the future as well as to the past.

A New Math for the CNS?

There is a growing idea in the world today that modern mathematics has failed to provide a proper framework to deal with problems of complexity that are encountered daily in the natural and social sciences. At meetings of economists, ecologists, engineers, physiologists, management consultants, and others we hear speakers announce that we need new mathematical constructs to deal with complex systems. Even so applied a publication as the *Wall Street Journal* has pointed out, "A search for new ideas to cope with today's troubles is intensifying. 'We need entire new ideas about what causes economic cycles and what to do about them,' says Norma Pace, economist for the American Paper Institute in New York."

It may seem surprising that so soon after the computer revolution and the great growth of mathematical awareness that accompanied it we are already facing a discontent over the power of mathematics to solve our problems. However, it is precisely the ability of the computer to push calculation so extensively that has revealed the limitations of the basic premises.

So many of the areas that we wish to deal with today focus not so much on individual events as on great numbers of interacting events. A collection of such elements is usually designated a system, and we are often interested in the global behavior when we know something about properties of the individual constituents. Dealing with such systems from a mathematical point of view involves a large number of simultaneous equations, either algebraic or differential. If these equations are linear (the unknowns occur independently as x, y, z, etc.), very general methods exist for obtaining solutions. However, if the equations are nonlinear (involve expressions such as x^2, xy, xz, etc.), then no general methods presently exist for generating satisfactory solutions. Most realistic modeling of situations encountered in the various systems' disciplines involves nonlinear equations to describe the interactions. Thus, in many areas of inquiry that bear very directly on "the human condition," progress is effectively blocked by an inability to formulate solvable models. As we have noted, one response to this frustration has been the suggestion that we need a new mathematics that can deal with complexity more effectively than can existing approaches.

The call for mathematical "revolution" occurred earlier in the computer age than may be generally realized, in a series of lectures prepared by John von Neumann in 1955 (published as *The Computer and the Brain* by Yale University Press in 1957). The poignant story behind this work is worth mentioning as an introduction to the ideas themselves. John von Neumann was one of the great thinkers of the 20th century in the area of mathematical foundations of science. In the 1940s and early 1950s he was one of the major developers of the concepts of software in the newly developing field of large, high-speed computers. Seeking a heuristic tool in developing programming, he turned his attention to neurology and the operation of the central nervous system. In 1955, while he was writing up his ideas on the relation of the computer to the human brain, in preparation for giving the Silliman lectures at Yale University, he became ill with cancer. During the terminal period of his life he worked on this material when he was able and the unfinished fragmentary manuscript was published posthumously. It is this work that contains the challenge to mathematicians that I shall discuss. His introduction notes, "I suspect that a deeper mathematical study of the nervous system will affect our understanding of the aspects of mathematics itself that are involved. In fact, it may alter the way we look on mathematics and logics proper."

Von Neumann then proceeded to discuss analogue and digital computers from the point of view of the basic operations ("four species of arithmetic: addition, subtraction, multiplication, division"), the hardware components, the concept of logical control, and the need of a special memory organ. Next he treated the need for high precision in digital computers, which follows from the fact that complex procedures require a very large number of steps, and errors are additive. Finally, he discussed the time requirements of different computer components.

Regarding the brain, von Neumann started with the statement that "the most immediate observation regarding the nervous system is that its functioning is *prima facie* digital." Although he recognized that the all-or-none view of the nerve impulse was an oversimplification, he nonetheless felt that the behavior is sufficiently binary to impose a basically digital logic on the system. He then compared computers and brains with respect to size of components, power requirements, and number of elements. This was followed by the startling conclusion that "the language of the brain is not the language of mathematics." He elaborates as follows: "Just as languages like Greek and Sanskrit are historical facts and not absolute logical necessities, it is only reasonable to assume that logics and mathematics are similarly historical, accidental forms of expression. They may have essential variants, i.e., they may exist in other forms than the ones to which we are accustomed. Indeed the nature of the central nervous system and the message systems that it transmits indicates positively that this is so. . . . Thus logics and mathematics in the central nervous system, when viewed as languages, must structurally be essentially different from those languages to which our common experience refers. Thus the outward forms of our mathematics are not absolutely relevant from the point of view of evaluating what the mathematical or logical language *truly* used by the central nervous system is."

The previous statements are so radical that we might totally ignore them if they did not emanate from an individual so universally respected as a scientist and mathematician. Von Neumann is suggesting that if we wish to understand a certain class of problems we must go out and create an entire new mathematics, different in kind from what we know today. There was, of course, a period of history when mathematical advances seemed to respond to the needs of physics; one can think of the names of Lagrange, Gauss, Laplace, Hamilton, and others. But the idea of a totally alternative mathematical language in response to the problems of biology seems to be calling for an even greater level of innovation than that associated with the great mathematicians listed above.

In any case, it is clear that in many areas of human endeavor we are finding mathematics less than equal to the problems that face us. With the researcher's optimism we assume a genius will appear to show us a new and better way of approaching these problems. However, we lack the algorithm to predict when that new development will occur, so that in the meantime we had better do the best we can with the old math. Genius is always a chancy thing, and it is generally best not to pause while we are waiting for it to come along.

Puzzle Solving In Science

The arrival in my mail of a new edition of *The Structure of Scientific Revolutions* by Thomas S. Kuhn (University of Chicago Press, 2nd Ed., 1973) led me to a reexamination of this enigmatic work. What was particularly intriguing is that the new copy had been sent by a friend who is a solid experimentalist, seldom given to philosophical speculations. Thus, my colleague's implied reaction induces further thoughts on the impact of this provocative book. My first copy was a gift from an undergraduate who wished to educate me. It has since been appropriated by my son, who uses it for interpretive reading in political science. I have also seen copies in a college bookstore shelved under English 10a.

Kuhn sets forth the thesis that science is structured both intellectually and socially by a series of paradigms, which he defines loosely as "universally recognized scientific achievements that for a time provide model problems and solutions to a community of practitioners." He further characterizes a paradigm as an "achievement sufficiently unprecedented to attract an enduring group of adherents away from competing modes of scientific activity." A paradigm is sufficiently open-ended so that it leaves problems for its practitioners to solve. By way of example are major paradigms such as Newtonian physics and Copernican astronomy. There are many lesser examples that define the sub-specialties of science. The general theory of macromolecular synthesis in molecular biology has been nicknamed "the dogma" in apparent recognition of its paradigm nature.

Kuhn argues that normal science is almost completely guided by its paradigms, and research is a strenuous effort to force nature into conceptual boxes supplied by professional education. He notes that our image of science comes from textbooks, "works both persuasive and pedagogic." The textbooks imply that additions to the subject come in a continuous way in an ongoing attempt to refine our view of the world. This, Kuhn maintains, is historically inaccurate, and major changes occur in more drastic and more radical confrontations between an existing conceptual structure and a challenging paradigm evoked by the failure of the present constructs to deal with the full domain of developing experimental results.

Persuasively, Kuhn argues by a number of examples that "an apparently arbitrary element compounded of personal and historical accident is always a formative ingredient of the beliefs espoused by a given scientific community at a given time." It is obvious that scientific views change and that individual influential scientists do much to mold the thinking of their time. Once a viewpoint is established the adherents have a vested interest in maintaining the viewpoint. I am reminded of my own days as a graduate student in biophysics when many of the botanists and zoologists were positively hostile to our small coterie of deviants. At the time I was totally unaware of the nature of the problem. Weren't we all seekers after the answers to the question "What is Life?' This book on the nature of scientific activity begins to make clearer the nature of a certain unfriendliness between groups of scientists.

Yet, Kuhn's view of the world is somehow jolting and offensive to the traditional scientist. After all, many of us feel that in spite of compromises with reality we are seeking after truth with a capital T. We regard ourselves as a community of scholars seeking to measure and understand the real world. It is disconcerting to be told we bear a relationship to a group of fraternities, each with its own rules, rituals, and doctrines. We are uncomfortable at the notion that we follow a manual of discipline that defines how far our thoughts can go. Thus, the first reaction to this book is tinged with annoyance.

The analysis of the scientific process then advances with a second ego-shattering notion that normal science is puzzle solving. Once the outlines of the paradigm have been established by the founders, it is maintained that subsequent adherents work out remaining problems within the conceptual outlines. A group of researchers accepts the same rules and standards for practice, and this apparent consensus is the necessary condition for normal science. Kuhn argues that "by focusing attention on a small range of relatively esoteric problems, the paradigm forces scientists to investigate some part of nature in a detail and depth that would otherwise be unimaginable." Again it is rather disconcerting to be told that our arduous, tiring, and often frustrating efforts are in the nature of puzzle solving. "Puzzle solver" may seem like a fair appraisal of someone else, but it is difficult to accept as a self-assessment. I think that in this regard scientists are not different from other professionals. It is relatively easy to see the repetitive, constrained paradigm behavior of one's peers; it is usually less apparent that one's own behavior falls into a similar category.

Paradigms within the Kuhn theory do not last forever. It may happen that data found in filling out the details are inconsistent with the framework. Such an event occurred when Roentgen noticed a barium platinocyanide screen glowing near his shielded cathode-ray apparatus. It may also happen that results in related fields affect the assumptions of a given specialized area. Thus, Planck's revised treatment of black body radiation in thermodynamics had its impact on the theory of the photoelectric effect. Other experimental findings or shifts in thought patterns may lead to the questioning of a method of approach, the rejection of a theory, or the proposal of a new viewpoint. The breakdown of Ptolemaic astronomy resulted from a very large accumulation of data that were more simply interpreted from another point of view. These major shifts in outlook are what Kuhn labels as "scientific revolutions." The group of researchers advocating the new viewpoint is in conflict with the established practitioners, and this leads to a struggle for the acceptance of ideas. Kuhn argues that "competition between segments of the scientific community is the only historical process that ever actually results in the rejection of one previously accepted theory and the adoption of another."

Thus, there is postulated a pattern of scientific progress. A way of thought becomes an established paradigm because in the view of the scientists involved it presents the most coherent framework for understanding some domain of nature. The paradigm provides the rules and norms for intensively investigating that domain in great detail. Those investigations plus related material from other disciplines provide the basis for rejecting the paradigm in its original form and proposing new theoretical constructs for understanding the field of study. The new viewpoint gathers its adherents and in time replaces the previous ideas to become the new paradigm. There is an almost dialectic flow to Kuhn's scenario.

The Structure of Scientific Revolutions is a curious blend of history of science, philosophy, and social psychology. It is another attack on the absolutist view of science that we inherited from the 19th century. It stresses that science is the activity of people who bring to their work all of their human frailties and limitations and form part of a social structure as well as an intellectual structure. It is a valuable book for developing professional humility. One suspects that the central theme has been carried too far. There is more continuity and asymptotic approach to philosophical understanding in science than Kuhn gives us credit for. It is hard to pick the appropriate point of compromise between the arrogance of an overly absolutist view of science and the know-nothingism of reducing all science to games people play. Also, science as we know it is such a new human endeavor it is perhaps premature for such sweeping historical generalizations. Never mind the criticisms, however, this is an exciting book that triggers young scholars into more probing thoughts about what they are doing. Who can ask for more?

Who Are the Standard Bearers?

It is fashionable these days to comment on and perhaps to lament about the low state of standards in the federal government. Not wishing to be totally swept away in this wave of cynicism, one finds it instructive to look at one branch of government specifically charged with the maintenance and development of national standards of measurement and their application; I refer to the National Bureau of Standards. This agency was established by an Act of Congress in 1901 and has continuously functioned as the national guardian and developer of scientific standards. At present the entire scientific, engineering, and medical community is deeply dependent on the activities of this organization. Not only such obvious areas as weights and measures, but pH measurements, x-ray safety requirements, radioactive counting standards, and many other less obvious examples all depend on activities of this scientific branch of the federal government. Some of the services in health-related activities can be seen in a partial listing of projects that includes medical thermometry, ultrasonic power measurements, microcalorimetry for biomedical assay, reference methods for the determination of lead in blood, corrosion of titanium in physiological solutions, and radiation standards in cancer therapy.

The type of role that the Bureau is capable of playing in

clinical chemistry can be seen in the following, taken from the 1972 report of the evaluation panel for the Analytical Chemistry Division.

> The Analytical Chemistry Division, as a group of long-term leaders in the field of certified chemicals and standard procedures for chemical analyses, can ably serve in the vital role of expert, impartial third party in the area of clinical chemistry.
>
> In the scientific community, especially among those closely associated with clinical chemistry, it is recognized that the design and manufacture of instruments for clinical analyses have been greatly deterred by the absence of certified (referee) methods. For some of the determinations most frequently requested by physicians, there are literally dozens of methods employed, all of unknown accuracy.
>
> Hence, the Analytical Chemistry Division, which recently made a modest start in the preparation of certified clinical chemicals (cholesterol and bilirubin), can greatly spur the clinical instrumentation industry by extending its capability to the certification of referee clinical chemical methods.

Maintaining scientific, engineering, and commercial standards is an absolutely vital yet often lonely area of scientific endeavor. The popular press and scientists themselves are most interested in the latest hot result on the forefront and seldom stop to think of the background structure necessary to make that result meaningful. There are few Nobel prizes likely to be given for extending a measurement by another significant figure or publishing a table of standard values. There has been a consistent lack of recognition of the importance of a strong standards foundation. As a result, the funding of many of the activities of the National Bureau of Standards has for the past several years been at or below the maintenance level. Important activities have been dropped because of job freezes, budget impoundments, and simple unwillingness of Congress to allocate more of the nation's resources to these functions.

Yet in spite of the difficulties outlined above, a visitor to the Bureau's campus at Gaithersburg, Maryland, cannot fail to be impressed by the dedication with which these scientists carry out their tasks. A certain élan exists along with a sense of seriousness about the job being carried out. These are a special breed of men and women who devote their professional lives to making science possible as a community activity.

It is a short step from the necessity of standards of measurement to a consideration of standards of integrity in the research community. A few recent and well-publicized examples of data-faking have sensitized all of us to the problem of honesty among scientists. Even beyond the ethical imperative to be honest there stands a structural imperative that makes science possible only within the context of the ideal of absolute honesty. All workers depend on a massive and complex literature to acquire the background information necessary for any project. We cannot individually repeat the experiments back to Pasteur, Franklin, or Galilei without so consuming ourselves in background that we are unable to move ahead. As long as that literature has a tradition of overwhelming honesty, it is a prime tool in research. Were it to become heavily contaminated with deliberate fraud, it would fragment the community of science and considerably decelerate the rate of progress. Were the element of fraud to become too large, the subject would totally collapse.

Scientists are not necessarily more honest than other persons, but the structure of the subject imposes a demand for adherence to a set of rules that keeps the structure intact. By analogy, musicians are not necessarily more cooperative than other people, yet the structure of the symphony makes an absolute demand on their playing in strict synchrony. If the amount of fraud in science is very small, the offending entries in the literature will eventually be found out and labeled as "unrepeatable." Not all unrepeatable entries are of course frauds. Nature is too complex to make life that easy for us. Nevertheless, the ethically neutral judgment of "unrepeatable" makes it possible to lump occasional lapses of ethics in with the vagaries of nature. However, if dishonesty were to rise to more than the tiniest fraction of the total effort such a solution would become impossible. Scientists are an international fraternity whose one sacred rule is "Thou shalt not bear false witness to nature."

Returning to the problems of nationally maintained standards, one begins to muse on the possibility of a National Bureau of Ethical and Political Standards. Envision for a moment a constant-temperature, constant-pressure room containing a completely honest politician—a treasure guarded like the platinum iridium standard kilogram. Other politicians are periodically brought in and checked against this standard. They are calibrated in units known as the standard Washington, which is the fraction of the time a politician tells the truth, minus 0.5, divided by 0.5. On this scale, complete honesty rates one Washington, a half-and-half liar gets a rating of zero, and a complete liar scores minus one. We have in public life seen some officeholders come precariously close to that minus-one level.

In another controlled environment room there resides a completely compassionate individual, the standard Lincoln. He is placed on one side of the compassionograph and national leaders, deans, and bureaucrats are brought in to be calibrated. A research group is trying to determine whether a high compassion index correlates with good or poor administrators. Moving down the hallway, we come to yet another comparison-testing situation. At the desk sits the national standard of political intelligence. He is rated by definition as 1.0000 Jeffersons. Across from him is the test subject and both are writing on large sheaves of paper. Although the ratings are highly classified, rumor has it that one well-known senator checked in at only 150 micro-Jeffersons.

Well, the dream is over. We do not and cannot reduce ethical behavior to a numerical basis. We can, however, hope that the national conscience will be aroused to where politicians will regard ethics as scientists regard honesty, a structural necessity for doing their jobs and maintaining the body politic.

From Aalto to Zwingli

A few years ago, upon opening a familiar journal, I encountered a book review of an encyclopaedia. The act of reviewing an entire work of this magnitude seemed at first sight to give fresh meaning to the word "chutzpah" and led me to Webster's *New World Dictionary* to find that noun defined as "shameless audacity." Research into less well-known and less reliable sources produced an account of the first translation of chutzpah into the Greek term "hubris." In one of the earliest interactions of Aramaic and Hellenic scholars in Alexandria, a debate took place as to the suitability of this translation. It was finally resolved that the subtle differences could be understood in the following aphorism: "When Prometheus stole fire from the gods, that was 'hubris'; when he tried to sell it back to them, that was 'chutzpah.' "

The previous remarks on audacity are meant to serve as an introduction and apology to a discussion (not, heaven help us, a review) of the 15th edition of *Encyclopaedia Britannica*. This version deserves special attention because it subdivides the universe of knowledge according to different principles than have been in use since the first edition in 1768. I also believe that given the relatively low price of small computers or telephone terminals, this is the last printing of a major encyclopaedia that will not have a computerized

221

cross-index. To establish credentials for the topic at hand let me note that the new edition has been in use in our family library for four years, standing next to the earlier one, which has been in place since 1964. By way of full disclosure, I must reveal being an unrepentant encyclopaediophile, a condition that seems congenital, but must have been acquired somewhat later than is implied by that term.

The major innovative feature of the new *Britannica* is splitting the 30-volume work into three sections: Propaedia, a one-volume introduction to all knowledge; Micropaedia, a 10-volume short-entry encyclopaedia; and Macropaedia, a 19-volume series of 4,207 long, scholarly articles covering the editor's view of the major topics of human learning. Micropaedia has extensive references to the articles in Macropaedia related to any given entry.

Propaedia is basically a long outline dividing the world of learning into 10 great divisions (Matter and Energy, The Earth, Life on Earth, Human Life, Human Society, Art, Technology, Religion, The History of Mankind, The Branches of Knowledge). This classification of knowledge goes down through six hierarchical levels, and each entry is liberally indexed to the appropriate articles in Macropaedia; thus this volume serves as a table of contents. Propaedia emerges as a work rather nice to own, but difficult to use. It is comforting to have the consensus view of what is known spread out "like a patient etherized upon a table." There is a conventional wisdom, and it is embodied in a broadly based classification system of this kind. One wonders, for example, how the average liberal arts education measures up to the 10 great divisions. I suspect it ignores six of them, for the classification is heavy on science and technology. It's also a little disconcerting to note, for example, that in the outline of all knowledge, sound-recording and -reproducing devices get the same amount of space as ethics or moral philosophy. Moderns may in fact be more interested in the acoustic fidelity than we are in the moral quality of that which reaches our ears. Such speculations can make Propaedia an endlessly fascinating volume.

Micropaedia probably gets more use than the other two sections. For the task of ready information retrieval I must confess I find this part of the new encyclopaedia less satisfactory than the older edition. A number of efforts to simultaneously look up items in the 14th and 15th editions leave me a fan of the earlier approach, and those volumes in our study probably get more use. However, every encyclopaedia has, of necessity, built-in obsolescence, and we must discipline ourselves to the later version to avoid being stuck in the past. There is a narrow line between venerable and old-fashioned.

In looking over the earlier edition, it is apparent that the most worn book in the collection is Volume 30, the index and atlas. Over the years that book has been the portal through which we passed on the way to answering a multitude of questions. The index tied the entire collection together in a way that neither Propaedia nor the extensive cross-referencing manages to accomplish. One individual's opinion affirms that there is no substitute for a good index.

Macropaedia has proven a superb topical encyclopaedia. The articles are often longer than one would desire, but given patience the sought-for information usually emerges, and the series has fulfilled its function. Nothing in the quality of material leads me to lessen my great respect for such an incredible undertaking as a major encyclopaedia like this one. I do believe, however, that there is an information access and retrieval problem which makes the latest *Britannica* a bit out of keeping with our age of information-handling technology—I wonder if it is too late for an index, or a small home computer or telephone access to a central computer index. Any of these would go a long way toward making the 15th edition a far more useful set of books.

As a final test of the new work I decided to see what I could find out about the concepts of hubris and chutzpah. Concerning the former, I now have a fine Micropaedial knowledge and can at a moment's lull in the conversation discourse on the arrogance of Xerxes in building a bridge of ships across the Hellespont. Sadly I must report that an

entry entitled "Chutzpah" is not to be found. Not that knowledge on the subject is unavailable; the 1916 edition of the
Funk and Wagnalls *Jewish Encyclopaedia* has an authoritative
article on the use of this Aramaic word in the Talmud. Perhaps the *Britannica* editors will, in their wisdom, include
"chutzpah" in the 16th edition. In any case, finding something missing in the 42 million words of the new encyclopaedia is really an act of hubris, or is it chutzpah?

Psychosclerosis

I remember Miss Martin as a rather short woman; tiny might be a more appropriate adjective. Yet we fourth-graders took her very seriously. I can't quite tell whether it was force of personality or the fact that her fiancé, who occasionally picked her up at school, was a local policeman, a strong, powerful-looking man. It was clearly a more innocent age.

In any case Miss Martin constantly stressed the importance of vocabulary. "When you come across a new word," she would say, "look it up in the dictionary and then use it at least three times until you are absolutely sure that you understand it." Homework assignments would often consist of finding a word and bringing in three sentences containing that new-found addition to our vocabulary. The teacher assured us that increasing word usage would bring great benefits in life not only in the G. W. Krieger Elementary School but in the wider world. Her ideas agreed with what I was taught at home and so the lesson stuck very firmly.

Therefore, two years ago when I encountered the word *propaedeutic*, the natural response was to go to the dictionary and find it defined as "A subject or study which forms an introduction to an art or science or to a more advanced subject generally." It could also be used as an adjective or in the plural. I was immediately taken by the utility of the term, since it embodied a concept I had often wished to express and yet had lacked the single appropriate word for its statement. Indeed, I was writing a volume on the physics underlying our understanding of energy exchange in biology and the title *Propaedeutics of Bioenergetics* suggested itself. Thus the dual objectives of titling the book and following Miss Martin's dictum could be accomplished.

A few nights later two scientist friends were over for dinner and I tested the newly generated title. To my great surprise, they became irate; they were hurt that I would use a word unknown to them. I was accused of offensive intellectual snobbery. Miss Martin had never prepared me for this response, and I must confess that the tenor of the dinner conversation considerably inhibited my enjoyment of the dessert and Irish coffee courses. The response was so intriguing that I tried the title on a number of other scientists, and their reactions covered the full range from boredom to disbelief.

At this juncture it became necessary to look into the background of my recently acquired word more seriously. It turns out to be a very venerable part of our cultural heritage. Immanuel Kant frequently employed *propaedeutic* in his writings, which provided the foundations of metaphysics for the next few centuries. You might say that Kant's *Critiques* are propaedeutic to modern philosophy. In addition, the editors of *Encyclopaedia Britannica* decided to call the first part of the completely reworked edition *Propaedia*. However, not withstanding the respectability of the concept, I did desire the sales of my forthcoming book to be greater than two or three copies, so, with apologies to Miss Martin, it was retitled *Foundations of Bioenergetics*.

Some people have on occasion detected in me a small stubborn streak, which asserted itself a few months later when I began a talk at a scientific conference by pointing out that the Mitchell hypothesis is propaedeutic to understanding energy transformation in living cells. The audience woke up, coughed considerably, and sputtered a bit. At least I finally had used the word once, and in public.

My next attempt occurred when in writing a magazine article I argued that certain 19th-century research was propaedeutic to modern physiology. The editors crossed out this novel bit of vocabulary and replaced it with "seminal." When queried, they replied that since none of the three co-workers in the office knew the meaning of the word it would be inappropriate to subject the readers to the experience. I lost that round, but kept wondering how the readers were going to increase their vocabularies beyond that of the editors if they never saw a word that the staff didn't already know.

Well, no experience should be so traumatic that we don't learn something from it, and my unpleasantries with *propaedeutic* leave me wondering why we condemn in adults that which we extol in children. Why do we assume that it is healthy for young ones to broaden their minds by new concepts and verbalizations while protecting their elders from any such liberalizing endeavors? Was William James correct when in 1890 he wrote, "It is well for the world that in most of us, by the age of thirty, the character has set like plaster, and will never soften again" (*Principles of Psychology*)? And, indeed, must the vocabulary harden as well as the character?

I would submit that we academics, intellectuals, editors, and the like are the victims of our own arrogance. We each assume that we, being educated, know everything about our native tongue that there is to know. Thus we are offended when someone uses an unfamiliar word and assume that the burden lies with the user rather than the reader. I would further submit that this is a very narrowing outlook and it constrains the mind from growing. We endorse mental ossification rather than accepting the basic humbling realization that "There are more things in heaven and earth, Horatio, / Than are dreamt of in your philosophy."

One of the most liberating events in my life occurred when I was able to stand up before a class and respond to a student's question with the reply, "I don't know the answer, but let's see what we can find for next time." We are all finite human beings; there is no disgrace in not knowing everything. There is, however, some problem with being unwilling to reach beyond what one knows to a broader, fuller reality.

A few years back, one of the slogans of the young was "never trust anyone over thirty." I wondered about the meaning of that phrase, but it is becoming clearer. If those over thirty are unwilling to learn even a new word, how much less willing will they be to accept a novel concept to respond to a changed world with adequate freshness and innovation. It is sad to think of all those minds about us setting like plaster, never to soften again. We cannot avoid the hardening of the arteries, but we are not forced to accept the hardening of our field of knowledge, a kind of psychosclerosis.

In any case, this has not been a totally negative experience; I have learned a new word and tried to use it three times. If I have only succeeded once, I hope Miss Martin will understand. And for those of you who didn't like *propaedeutic*, I strongly advise you to be more receptive to new knowledge lest you spend your declining years sitting around doing nothing but engaging in omphaloskepsis.

PART VII

Drinking Hemlock and Other Nutritional Matters—THOUGHTS ON WHAT WE KNOW

Drinking Hemlock and
Other Nutritional Matters

It was a rather dark, bleak morning, and after rising early I thought it appropriate to turn on the television and communicate, unidirectionally to be sure, with the outside world. There to my great surprise was a famous movie star of a few years back discoursing on the evils of sugar. The former Hollywood idol was vehement in her denunciation of this hexose dimer particularly in its purified and crystallized form. She denounced it as an "unnatural food," an epithet that may well have bruised the egos of the photosynthesizing cane and beet plants. The mental image evoked was that of a solemn judge sentencing someone in perpetuity for an "unnatural act." In no time at all this great lady had me caught up in her crusade, and I kept muttering "hate sucrose" as I prepared an unnatural extract of coffee beans and dropped in a highly synthetic saccharin tablet.

A few minutes later, when the veil of sleep had lifted and the uncertainty of reason had replaced the assuredness of emotion, I began to wonder where my cinema heroine had acquired such self-righteous certainty about biochemical and nutritional matters that have eluded my colleagues for years. Perhaps all this messy experimental work of grinding and extracting tissue and otherwise mucking about the laboratory is not the shortest road to truth at all, and we of the dirty white lab coat crowd are missing some mysterious pathway whereby true nutritional knowledge comes with blinding insight and transforms the lives of the faithful.

All of this recalled a frequent, painful experience that haunts biomedical scientists like a recurring nightmare. One is at a cocktail party or other social gathering where someone appears in the crowd and begins an oratorical declamation on Good Nutrition. The "facts" being set forth are often inconsistent with everything one knows about metabolic pathways, cell and organ physiology, enzymology, and common sense. If the listener is so bold as to raise the question, "How do you know that?", he or she is greeted with a look that must have faced Columbus when he queried, "How do you know that the world is flat?"

Nutrition seems to be like politics; everyone is an expert. It would appear that to the general public years of education are as naught compared to knowledge somehow painlessly available to everyone, regardless of his familiarity with innumerable facts and theories that constitute a complex discipline.

The situation described is by no means confined to the choice of foods, and I certainly feel ill prepared to get involved in the sucrose controversy. Nevertheless, the field of nutrition is a good example of the many areas where we are constantly subjected to a host of dogmatic statements, some of which are true, some of which are false, and many of which are indeterminate. The response to each of these assertions should be the query, "How do you know that what you are saying is indeed a statement of fact?" At this level of question, I believe our educational system has been a total failure.

Asking how we know the things that we know is part of the philosophical discipline of epistemology, the theory of knowledge, which is usually taught in upper-level and graduate philosophy courses and is therefore restricted to a small group of college students. But can there be any study that is more basic to education? Should not every high school graduate be prepared to cope with the many incorrect and misleading assertions that come his way every day? On the surface it seems strange that acquiring skills in assessing the validity of statements is not a core feature of the school curriculum.

Education, as conceived at present, is largely a matter of transferring subject matter from teacher to student, and uncertainty is usually settled by appeal to authority, the teacher, a textbook, or an encyclopedia. The methodological issue of how knowledge is obtained is rarely mentioned. Thus one of the most important analytical tools that an educated individual should possess is ignored. This is not to argue against the transfer of information but rather to assert that by itself it is insufficient protection in a real world containing demagogues and all kinds of charlatans and hucksters who have a free reign because almost no one is asking the appropriate questions.

On the issue of sorting out reality, most holders of doctoral degrees are almost as naive as grade-school graduates, and all manner of academic disciplines also expend effort on statements that would be quickly discarded if epistemological criteria were invoked. This takes us back briefly to the subject of nutrition, where methodological problems make it very difficult to obtain even pragmatically useful information. Statements are made on the basis of averaging over populations when we have no idea of the distribution functions that go into forming the averages. The impossibility of large-scale experiments with people requires extrapolation of animal or small-scale human determinations over ranges where the correctness of the extrapolation procedure is unknown. Nutrition is thus beset with difficulties that are clearly of an epistemological nature and, until these are resolved, careful scientists will be confined to very limited statements. Dogmatic assertions will remain the province of cocktail party orators.

The problem of why the theory of knowledge is not taught in the schools is relatively easy to see. Epistemology is, after all, a dangerous subject. If we start to question the validity of statements, then the teachers themselves come under question. All assertions about education, established forms of religion, government, and social mores will also be subject to justification on the grounds of how they are known to be true. For parents and teachers who have not been through the experience of exploring how we determine facts, it would be unnerving to have their children continuously questioning the roots of knowledge. Inquiry is indeed a challenge to the acceptance of things as they are.

To realize the threat to established ways that is perceived in the type of analysis we are discussing, we need to go back to ancient Athens, where the philosopher Socrates taught his young followers by the technique of questioning everything and seeking answers. As Will Durant has noted, "he went about prying into the human soul, uncovering assumptions and questioning certainties." This has come to be known as the Socratic method. The citizens of the Greek city-state condemned the inquiring teacher to death by poisoning with hemlock. One of the most serious charges against him was "corrupting the young." The fate of the first propounder of the Theory of Knowledge has perhaps served as a warning to keep the subject out of the school system.

There is still an objection that it is dangerous to teach the art and science of inquiry to the young; I would submit that it is more dangerous not to teach it to them, thus leaving them vulnerable to the quacks and phonies who now add mass communication to their bag of tricks. If we believe that rationality will lead the way to the solution of problems, then we must start by making the examination of what is "real" a part of everyone's thought. If challenging young people are a nuisance, think of how much more of a menace is presented by young people marching off in lock step and never questioning where they are going.

The solution seems clear. When we return education to the basics of reading, writing, and 'rithmetic, we should add a fourth R, "reality." Starting at the first grade and continuing through graduate training we must see that students become sensitized to the meaning of what is said and the realization of how valid knowledge is established. If this seems radical, it is. Drinking hemlock may be less painful than swallowing some of the drivel that comes over the TV set every day.

Food for Thought

Nutrition, on first inspection, appears to be such a down-to-earth discipline that one has a moment of doubt about viewing it in relation to abstract subjects like philosophy of knowledge and theory of probability. But the issue is forced upon us. Someone out there is very concerned about what we eat and shouts at us over television, puts pictures on the cover of consumer magazines, clutters our mail, and sets a rash of titles in front of us at the bookstore. We are advised to avoid sucrose and saturated fatty acids. We are instructed both ways on vitamin C, and we are placed in unholy terror of the dread cholesterol.

I must confess that this "hate sterol" campaign is beginning to get to me. After all, we are all about one half of one percent cholesterol, and this self-negation seems a bit pathological. Cholesterol provides me with cell membrane material and keeps my organelle envelopes in an adequately fluid state. Besides, I dislike getting angry at molecules that occupy such an important position on the metabolic chart of animals; it seems an assault on nature. Actually, when I begin to think about cholesterol as the precursor of both male and female sex hormones, I can get downright romantic about this substance. All of which serves to introduce my theme. How do I know that any or all of those people out there advising me on eating know what they are talking about? Once that awesome question is raised, we move into the philosophical domain of the theory of knowledge, epistemology.

Let us start by trying to state the basic task of nutritional science, which is to formulate a diet over the lifetime of an individual that will optimize health, well-being, and longevity. This calls for providing the necessary chemical components in the right proportion and avoiding or minimizing toxic substances. Such a protocol assumes: first, that we know what is essential; second, that the quantitative requirements are uniform enough to be meaningful for a population; third, that deleterious substances have been identified; and fourth, that the foods and toxins can be separated. This separation is often complicated by the observation that a nutrient at one level of intake can be a poison at a higher level, as in hypervitaminosis. The first two assumptions bring into focus the kinds of experiments that establish requirements or nutritional values of foods. These are carried out on rats or on humans and suffer from a number of methodological difficulties. The mapping from rodents to *Homo sapiens* is not precise. This is often presumed to be answered by picking strains of rats that resemble humans with respect to known requirements, a procedure that guarantees nothing with respect to unknown requirements. The difference in life-style between rats and humans simply cannot be neglected. The experiments themselves are long-term, costly, and often difficult to interpret. Discovery of subtle effects requires, for statistical validity, a far larger animal population than is usually practical.

A flagrant example of the problems of rat experimentation is found in the May 1976 *Consumer Reports*, where breads are "nutritionally compared." Rats in groups of six were tested for 16 weeks on a sole diet of one brand of bread per group. Without any discussion of primary data or statistical significance, the breads were rated for quality. Such ill-conceived and inadequately reported experiments are supposed to guide the consumer.

Human experiments are invariably short-term with respect to life span and also suffer from the problem of too small a population. It is relatively easy to spot well-defined relations such as vitamin C and scurvy, but it is exceedingly difficult to evaluate less dramatic correlations. All of the problems met with in nutrient requirement experiments are encountered in toxicity studies. Acute toxic effects can be established, but the long-term nexus between various food additives and incidence of disease is extremely hard to come by.

A further methodological difficulty must be considered. Various nutrients and toxins have combinations of antagonistic and synergistic effects that render the response to single factors of limited usefulness. In the language of the systems theorist, we are dealing with a multiple input system, in which the signals are integrated in some manner that, at best, is incompletely understood. Having to consider interactions, even at the elementary level of two at a time, intensifies the problems, since for every N possibilities there are $N(N-1)/2$ binary combinations to deal with. These difficulties are a result of the complexity of the system and cannot be avoided by simple devices of analysis.

Some additional information may be available from evolutionary considerations. If we study the foods that humans have eaten during the development of the species as determined by paleontological investigations, we have a long-term, if imprecise, set of data that might yield some useful clues. In addition, the diets of our primate relatives may contain further information. In some ways we know a good deal more about the nutritional status of domestic animals than about humans. The information is of immediate economic value in agriculture, and the organisms are more cooperative with the experimenters.

All of the preceding adds up to the fact that most of what we are told about nutrition is neither true nor false; it is indeterminate. The experiments that have been carried out are often inadequate to develop conclusions within the ground rules of probability and statistics and the accepted notions of scientific verification. The information available to diet planners consists of: a small body of universally accepted results such as the pathways of intermediate metabolism, a set of direct minimum requirements to avoid dietary deficiencies, data on toxic substances and levels of acute toxicity, and a very large body of results—many of which do not measure up to the minimum standards of statistical acceptability.

Given the Babel of information and misinformation, the question emerges: Is there a rational method to decide what to eat? To seek an answer, we must go back to a two-hundred-year-old idea that is usually referred to as "Laplace's Principle of Insufficient Reason" and states that events are to be assigned equal probabilities if there is no reason to think otherwise. E. T. Jaynes has recently extended the principle to state that "in making inferences on the basis of partial information, we must use that probability-distribution that has the maximum (*informational*) entropy . . . maximally noncommittal with regard to missing information subject to what is known." Without going into a detailed computer program, we can infer that the most rational aliment will be obtained by fulfilling known nutrient requirements, minimizing intake of proven toxic substances, and randomizing everything else. In other words, the most variable diet that fulfills known constraints is the diet of choice. The clue is variation: use different foods, change brands frequently, and never use the same recipe twice. Bring on the escargot, candied grasshoppers, roast guinea pig, alfalfa sprouts, macadamia nuts, falafel, turnip greens, squid, mare's milk, and seaweed salad. Minimize the intake of packaged foods, since they contain additives of unknown toxicity, and where they are unavoidable, keep varying the suppliers according to additives.

Beyond our personal diets, it is clear that the sciences of nutrition and toxicology require deep methodological reexamination. It has been my impression that these sciences have been largely underemphasized in medical school curricula, so that the individuals whose advice is most sought about foods, the practitioners, are inadequately trained and often unfamiliar with the full range of theoretical issues involved. The first step to remedy these problems consists of a probing examination of the methods of obtaining valid results within these fields of study. The second step would seem to be an upgrading of diet-related disciplines by medical faculties. Whatever else man may be, he is a mechanism that responds to a lifetime of inputs with a series of outputs. Those inputs that arrive through the gastrointestinal tract describe a large part of the human experience. At present they are of two categories: normal biochemicals that have been part of the human milieu for millenia and new organic compounds that have only recently been synthesized. Before we ingest any large quantity of these organics, it would be nice if we knew what we were doing. Nutrition could be a major field of medicine—and a most positive one—which aims at promoting good health rather than responding to system failures.

Hiding in the Hammond Report

The Hammond Report on "Smoking in Relation to Mortality and Morbidity" was the documentary result of a prodigious amount of data gathering, involving at the outset 1,078,894 subjects and 68,116 volunteer workers. Presented by Hammond to the American Medical Association meeting in Portland, Oregon, on December 4, 1963, the report was based on the records of 422,094 men who were traced for three years. Its overall conclusion is now emblazoned on every pack of cigarettes and was summed up in these words: "Death rates were found . . . to be far higher in cigarette smokers than in men who did not smoke cigarettes. . . ."

The study was designed to find the relation between cigarette smoking and mortality and the results were clear, but a statistical study of this magnitude may also be expected to produce unanticipated results—results that tell us something more about the human condition. And, indeed, an examination of the data reveals many other facets, some totally unrelated to the smoking problem.

In attempting to cover a wide range of possibilities, Hammond looked at mortality as a function of a large number of factors other than smoking. These were summed up in Table 7, "Age-Standardized Death Rates Per 100,000 Man Years for Various Groups of Men, Aged 40 to 69." In our quest for factors other than cigarettes, we will examine data for nonsmokers, and begin with a dietary correlation.

Subgroup	Age-Standardized Death Rate
No fried foods eaten	1,208
Fried food 1-2 times a week	1,004
Fried food 3-4 times a week	642
Fried food 5-9 times a week	781
Fried food 10-14 times a week	722
Fried food 15+ times a week	702

The results are as surprising as they are ambiguous: the death rate decreases as the amount of fried food increases. After years of being warned against fried foods by nutritionists, we find that this massive statistical study demands that we eat substantial amounts of fried foods in a quest for longevity. Yet we cannot ignore this result any more than we can ignore the surgeon general's warning that "cigarette smoking is dangerous to your health." Since the fried foods must contain many of the lipids that are regarded as being bad for the arteries, the result is even more counterintuitive.

Another, even sharper, conclusion appears when mortality is viewed as a function of sleeping patterns. The data for nonsmokers are as follows:

Average Hours a Night Sleep	Age-Standardized Death Rate
Less than 5	2,029
5	1,121
6	805
7	626
8	813
9	967
10 and greater	1,898

The results are clear, but the meaning is obscure. Either sleeping less than six or more than nine hours a night is extremely unhealthy or sleep habits are a diagnostic tool of major significance. In any case, the raw data demand an explanation. Either the alarm clock is an underutilized therapeutic tool or the medical profession is paying insufficient attention to an easily obtained bit of information.

When mortality is examined as a function of education, a clear trend can be perceived. Again looking only at non‑smokers, we see:

Education	Age-Standardized Death Rate
Grammar school or less	945
Some high school	864
High school graduate	766
Some college	755
College graduate	676

Now *there* is a result that should be posted over the ivy on the gates of every college to reverse the declining trend in admissions. Education is good for you in more ways than one.

Next, the correlation of height with mortality provides an interesting possible clue into the genetics of lifespan. Again the data are for the nonsmoker.

Height of Subject in Inches	Age-Standardized Death Rate
Under 66	1,065
66-67	815
68-69	806
70-71	784
72-73	687
74 and over	735

The implication of this table is that taller is healthier, at least up to six foot one inch.

The final example we shall choose deals with marital status and provides important clues to the sociology of this aspect of life. First examine the data.

Age-Standardized Death Rate

	Nonsmokers	*Cigarettes 20+ a day*
Single	1,074	2,567
Married	796	1,560
Widowed	1,396	2,570
Divorced	1,420	2,675

This table provides a real support to defenders of marriage in a world that is beginning to take the institution lightly. Among the nonsmokers, perennial bachelors are dying 30% faster than married men while widowers and divorced men are dying at almost twice the married rate. The message for men is clear: be nice to your wife and if you still happen to lose her, go out and get another one as fast as possible to minimize the time you spend in the high-risk category. Looking at the smokers' column along with the other data we note that being divorced and a nonsmoker is slightly less dangerous than smoking a pack or more a day and staying married. If a man's marriage is driving him to heavy smoking he has a delicate statistical decision to make.

The problem we face is what sense can we make out of all of this statistical material. We have to take it very seriously. It was obtained in an honest, systematic, and statistically significant study. The fact that we were not asking for these answers makes them even more striking because they are relatively free of bias in the way the questions were posed. We must seriously question our views on fried foods, ask where those views come from, and ascertain if they are as solidly based as are the results in this study. The extreme sensitivity of death rate to sleeping patterns calls for the most extended research on the physiology of sleep. We cannot ignore a parameter that appears to be associated with alterations of two or three in the overall death rate.

It is perhaps odd that these results have been around so long with little attempt to sort them out. In a study of the magnitude of the Hammond Report there are doubtless other surprises waiting to be found. Now that the smoke has cleared it is well to go back to this document and seek for the fascinating results that are hiding within the mass of numbers.

Dull Realities

Edgar Allan Poe, master of the horror story, once turned his attention to ordinary science and penned a sonnet, which begins:

> *Science! true daughter of Old Time thou art!*
> *Who alterest all things with thy peering eyes*
> *Why preyest thou thus on the poet's heart,*
> *Vulture, whose wings are dull realities.*

Dull realities were indeed detested by Poe, whose vivid writings depicted a world of demons, evils, miasmas, and even "The Masque of the Red Death." Poe did not create these terrors de novo out of his fertile imagination but relied on a tradition of fear as ancient as humankind itself, indeed perhaps older. For the countless adults and children who have, through the ages, experienced the horrors of unknown and unseen presences, the question of their reality is an abstraction of only academic importance. No, Poe did not invent terror; his genius, however, did find the words and cadence to "express the inexpressible."

The rise of science, the rationalism of the encyclopedists, and the hard-nosed scientific materialism of the 19th century did much to reduce the world populations of demons, phantoms, apparitions, bogeymen, and other nameless and shapeless frights. But these creatures of the darker side of our brains are by no means endangered species, for fear is much a part of the human mind, and no amount of superficial explanation will wipe away the awful uncertainties deep in the psyche. Indeed, under the guise of science, many dreads of yesteryear have returned to haunt a new generation whose extensive technology offers little protection from such primordial facets of our being.

A more systematic nomenclature now describes those haunting presences, and psychologists define a phobia as an excessive and unreasonable fear of some object or situation. One of the more scientifically rooted of these anxieties is the fear of contamination with germs. This dread bears the name *mysophobia* or *bacteriophobia*.

Abhorrence of infectious diseases was of course preeminent in Poe's time and is chronicled in "The Masque of the Red Death" and in "Shadow—A Parable," where it is noted that "far and wide, over sea and land, the black wings of Pestilence were spread abroad." The "Sphinx" begins, "During the dread reign of cholera in New York . . ." These fears could hardly be classed as phobias, since plagues struck regularly, and the number of victims was large. Indeed, Mr. Poe himself set forth a beautifully precise distinction between rational and irrational fear. In describing a friend, he noted:

> His richly philosophical intellect was not at any time affected by unrealities. To the substance of terror, he was sufficiently alive, but of its shadow he had no appreciation.

Bacteriophobia, whose focus is actually on the microorganisms of disease, would not have been possible in the world of our horror story writer, since he came to his grotesque end in 1849, several years before Koch and Pasteur demonstrated the germ theory of disease in such a convincing fashion. The unknown terror of earlier days became replaced by known ones, tiny demons who, without a will of their own, are carried on air, water, or a lover's lips to become implanted on our bodies, grow in our entrails, and strike us down.

It is now over 100 years since bacteriology became a science. Public health measures, antibiotics, and supportative therapy have considerably reduced the dangers of microorganisms to man, yet the phobias associated with these organisms persist, and one has the feeling they may even be on the increase. Two types can be discerned: one an individual condition and the second a broader-based feeling which finds its way into mass communication and governmental policies. The private phobias are found in casebooks of psychologists, which describe men and women who spend entire days in washing their hands or in breathing through filters. The public mysophobia is less well documented although it is clearly exploited by advertisements for soaps, disinfectants, mouthwashes, and similar over-the-counter products.

The depth of public reaction to microorganisms was brought home during a recent New York City power outage. While not in the city, I was within radio range and listened to the unfolding tale of looting, rioting, and arson. These reports were punctuated by frequently repeated warnings about the dangers to life and limb associated with eating food whose temperature had risen some degrees above the normal operating range of the refrigerator or freezer. To a biophysicist these very strong statements seemed misplaced from both bio and physical points of view. From a thermodynamic perspective the sides, top, and bottom of a refrigerator or freezer come reasonably close to the ideal of rigid adiabatic walls, barriers to the flow of heat. Since power was unlikely to be out for more than 12 to 18 hours, the best advice from thermal physics would be: "Don't open the refrigerator or freezer door." The nonconducting properties of the cabinet and the high specific heat of frozen foods would then tend to keep the temperature low.

The major biological factors seemed to be the microbial

flora of frozen foods and the growth rates of these organisms as a function of temperature. Since we could eliminate the possible increase, in food materials, of viruses pathogenic to humans, we were able to focus attention on bacteria and fungi. At this point the distinction had to be drawn between refrigerated and frozen items that were to be cooked and those that were to be simply eaten. In the former case, high temperature would destroy the cells, and we needed only to be concerned about heat-stable toxins. While such chemicals are capable of causing discomfort, they are not life threatening, and not much poison is likely to be produced in a few hours far below the optimum growth temperatures. Thus there seemed to be little reason for concern about food that was eventually to be broiled, roasted, or boiled, and a simple statement would have sufficed dealing with foodstuffs that were to be eaten uncooked.

The warnings given on the radio, however, were of such a severe character that they might well have led scientifically untrained listeners to discard vast quantities of perfectly good nutrient brought into suspicion by a hysterical public mysophobia. One had visions of two million listeners each throwing out fifty dollars' worth of the contents of their refrigerators and freezers, thus imposing an additional one hundred million dollars' worth of loss to the existing catastrophe.

There is a well known and persistent myth that frozen foods cannot be thawed and refrozen without becoming highly toxic. This doctrine in its simple form is patently false; one of the classical methods of inactivating bacteria is repeated freezing and thawing. Yet in spite of its falsity it has become a well-established dogma known to every cook and occasionally taught in our institutions of higher learning. Indeed even my own dear mother, who was very proud of my education, was unwilling to accept my academic credentials when I assured her that thawed and refrozen meat was perfectly edible. So the popularity of the *no refreezing* dictum reinforced the mysophobia, and therefore New Yorkers sat in the dark shivering in fear of the unseen creatures lurking in the freezer and converting their hard-earned groceries into offal.

The thought of the economic hardship as well as the psychological effects of the publics' being subjected to repeated harsh warnings about food eventually, after 15 hours had passed, led me to phone CBS radio in New York to discuss the dull realities. Alas, the newsroom after a power failure is no place for calm reflection, and my pleadings were lost on the nervous party at the other end who unceremoniously hung up on me. Therefore, I want to take this opportunity to alert the news media to the existence of bacteriophobia lest their phantasmagoric conceptions of food microorganisms cause us great problems when the next crisis rolls around.

High in the Andes

What circumstances would lead certain otherwise genteel people to spend an entire evening talking about diarrhea? To orient our approach to this somewhat loosely construed question, we must move in place to Quito, Ecuador, and in time to a few months ago. We go down a narrow set of stairs into one of two small dining rooms of Las Cuevas de Luis Candelas, one of the favorite eating spots of the old city. Ten visitors from the United States are seated at a long table and engaged in a lively discussion. The issue before them is whether or not to order the house specialty cocktail, since it comes complete with ice made from the dreaded local water. The diners are newly arrived in South America, and no one has yet faced the discomfort of gastrointestinal disorders. The scatological turn of the conversation is conditioned not so much by the perversity of these people as by the pervasiveness of the tales told to individual members of the dinner party by friends and associates who had passed through this part of the world or had this part of the world pass through them.

Such talk must frequently occur among wayfarers to the underchlorinated regions of the globe. In principle, what distinguished this exchange of opinion from thousands of others were the formal credentials of the participants. There were two certified pathologists, one nonpracticing MD who was a cell physiologist, two PhDs engaged in molecular biology research, one research assistant in biochemistry, and one nurse. The three nonprofessionals at the table were all experienced travelers who had been to a number of countries around the world. The curricula vitae of the professionals listed papers on bacteria, animal viruses, mycoplasma, and various assorted flora and fauna. All in all, one would have judged that this was a consortium well qualified to deal with the eating problems of tourists. The thrust of the conversation turned to considering which local foods were safe and which unsafe to eat. Hard science was mixed with anecdotal material. Medical microbiology was mixed with mysticism. Aspects of the dialogue were sufficiently timeless that they could have been echoes of Francisco Pizarro arguing with his brother Hernando about what the conquistadors should and should not eat. The immediate problem was solved by ordering the special cocktail, *sin hielo*, without ice. (I must report that it is not one of the world's great drinks in its room temperature state.) The original discussion of the etiology of digestive tract disease continued on through the evening and so it came to pass that ten nice people spent the next three hours dealing with matters not usually regarded as dinner table talk. In all fairness to Luis Yepez Baca, the restaurant owner, it should be reported that the food and wine were excellent and reasonably priced, so that the topic of the evening did not deter the discussants from having a most pleasant meal.

But now it must be noted that the erudition and experience of this company, which included five professors from some of the most prestigious universities and medical schools of the U.S., produced no consensus, nor even a sense of the meeting of the minds. Some, it turned out, used local ice, some did not. Some ate local salads, some did not. Some ate local fruits, some did not. Some brushed their teeth with beer or bottled water and ate only cooked foods, well done. And in the end, some got diarrhea, some did not. There was little or no correlation among all of these factors. Before the reader thinks too poorly of this collection of experts, let him ponder such questions as what is the inactivation time of encysted ameba at 0°C in a 20% ethanol solution, or what fraction of salmonella is killed by one freeze-thaw cycle? The question that emerges is that if the reported conclusions were the best that a team of sophisticated medical professionals could come up with, what about a poor layman, who is not so trained? Where does one get good, simple public health information about foreign countries? The first moral is that knowledge is hard to come by and wisdom is even scarcer.

We now move ahead in time to a week later. The same tourists, now considerably more worldly wise, are gathered for drinks at a hotel in La Paz, Bolivia. Once again the conversation takes on a clinical tone, as those in the party who are suffering from "high altitude sickness" are reluctant to risk the Scotch, duty-free though it may be. The obvious cure for altitude sickness is within their grasp, for at all street markets the local Indians are constantly chewing coca leaves to counter the effects of anoxia. And sure enough, as if by magic, a small plastic bag of coca leaves appears in their midst. But wait, a variety of medical questions arise. First, and most easily disposed of, are the worries about acute toxicity and development of dependence. No one is going to chew that much of the dirty-looking leaves. Then the more serious problem—how about the micro-inhabitants of the leaves and their effects on the already partially abused digestive systems? This issue was resolved by soaking the leaves in Halazone solution. An additional difficulty emerged from an inspection of the group's one source of written information on coca leaves, a paperback archaeology book that stated that the Indians chewed the leaves with calcinated sea shells. Now for the triumph of the intellect. It was reasoned that the effect of the calcinated shells could only have been to make the leaves more alkaline. It was also quickly discovered who carried the remedy on his person. Thus the Halazone-washed leaves were chewed along with antacid tablets by four of the daredevils in the crowd. Others spurned this trafficking with drugs as they sat calmly and sipped their alcohol. Truly, the whole experience was a tribute to contemporary science. The second moral is that whereas knowledge may fail, ingenuity knows no limit.

In the process of locating the antacid, the chance presented itself to make a catalogue of the pharmaceuticals being carried about the Andes by these medical scientists. At this point it was resolved to formulate a traveler's pharmacopoeia so that the accumulated if somewhat dubious wisdom of the group could be made available to the waiting world. Excluded from consideration were drugs carried by specific individuals for use in the management of previously existing conditions. The remaining list is a pharmacological Baedeker of western Latin America. The names listed are as given and no attempt was made to edit or reduce the list to generic names.

Codeine

Bufferin

Fiorinal

Aspirin

Coca leaves

Cafergot

Noxzema

Lanolin

Rubbing alcohol

Baby oil

Petroleum jelly

Desitin

Synalar

Aristocort

Compazine

Librium

Maalox

Tums

Parepectolin

Pepto-Bismol

Lomotil

Terpin hydrate

Coricidin

Sucrets

Otrivin

Luden's menthol
   cough drops

Actifed

Benadryl

Chloromycetin capsules

Bacitracin-neomycin
   ointment

Penicillin

Erythromycin

Tincture of iodine

Halazone

Tetracycline
Norflex
Coramine ampules
Dramamine
Chloroquine phosphate

Triethanolamine
Benzalkonium chloride
Tinactin
Desenex

The next and last chapter of this Hippocratic history occurred a week later, at Iquitos, Peru, where the same in-
trepid adventurers were seated after dinner on the veranda
of a lodge overlooking a tributary of the Amazon River. This
evening was a genuine postman's holiday, and everyone was
involved in a long, clinically oriented discussion. Seated
nearby was the cook, a young man who understood almost
no English, but enough, apparently, for him to realize that
he was being edified by an erudite medical seminar. In fact,
he responded in a perhaps predictable way by complaining
of a stomachache. Since he had prepared the meal a few
hours earlier, reports of his illness were alarming. One of
the participants, who was not an MD and was therefore uninfluenced by Peruvian malpractice law, produced a non-
prescription remedy from the traveling medicine kit. This
was gratefully consumed by the cook and all pulled up their
mosquito netting and went warily to sleep. The following
morning the cook and everyone else appeared happy and
healthy. Once again, modern medicine had emerged trium-
phant. The third and final moral is never underestimate the
power of positive thinking.

Lox et Veritas

I can remember several years ago sitting across the desk from a friend, a physician and a man of wisdom. In his soft voice and gentle Austrian accent he said, "I believe that salt is the most dangerous food additive in the American diet." We explored the subject further and although I followed the thrust of the argument, I felt that his thinking was heavily influenced by the fact that his wife was suffering from severe pulmonary edema. He reasoned that, since between 15% and 20% of the population were hypertensive and numerous others had fluid retention problems, the constant high level of sodium in the diet constituted a load that materially decreased the lifespan. Foods were heavily laced with sodium chloride, sodium bicarbonate, sodium nitrate, monosodium glutamate, sodium propionate, and a wide range of other sodium salts. He judged that this added sodium was unnatural, overloaded the physiological control systems, and frequently led to clinically recognizable symptoms.

That conversation was recently recalled to me by a television commercial urging a checkup for hypertension that was juxtaposed with commercials for foods with high added sodium. In an effort to reexamine the question of sodium content I went back to Agriculture Handbook No. 8—*Composition of Foods*, edited by B. K. Watt and A. L. Merrill and issued by the U.S. Department of Agriculture. This volume is a fount of all kinds of information on foods and at under five dollars is one of the great bargains for those interested in nutrition. Sodium values are listed in milligrams per 100 grams of edible portion, and 2,483 foods are listed, ranging from abalone to zwieback. For a select group of natural foods we find the following sodium values:

Apples	1	Cream—light	43
Apricots	1	Eggs	122
Asparagus	2	Grapefruit	1
Bananas	1	Grapes	0.4
Beans	19	Herring	74
Beef	65	Lamb	75
Beets	60	Lettuce	9
Blueberries	1	Milk	50
Bluefish	74	Mushrooms	15
Broccoli	15	Onions	10
Carp	50	Peanuts	5
Cashew nuts	15	Pork	70
Chicken	50	Potatoes	4
Corn	0.7	Wheat	3

These represent pretty good average values over the range of fresh foods.

Next let us examine some prepared materials, a group of nutrients whose mineral composition is in part or wholly under the control of the manufacturer.

Asparagus—canned	236	Bacon—cooked	1,021
Baby food—oatmeal	437	Beans—canned	236
Baby food—vegetable with chicken	307	Biscuit mix	1,300
		Bran flakes	925
Bread	500-600	Noodles	5
Cake	200-400	Olives	2,400
Chocolate—bittersweet	615	Pancakes	425
Cheese—American	1,136	Peanut butter	607

Cornflakes	1,005	Peas—canned	236
Corned beef—cooked	1,740	Pickles—dill	1,428
Crab—canned	1,000	Pie—apple	301
Doughnuts	501	Pizza	702
Gelatin dessert powder	318	Salad dressing	500-1,300
Jams and preserves	12	Frankfurters	1,100
Macaroni and cheese—		Soup—chicken	382
canned	304	Tuna—canned in oil	800
Margarine	987	Waffle mix	1,029

The difference between prepared and natural foods is the difference between a daily intake of under 800 milligrams of sodium and a daily intake of about 10 grams of sodium. Modern methods of food preparation clearly impose a heavy burden on the physiological sodium-processing system. There seems little reason to doubt that the present sodium intake is far removed from the diets of our immediate evolutionary forebears. It is therefore interesting to focus attention next on a group of very high-sodium foods.

Bacon—Canadian cooked	2,555	Olives	2,400
Beef—dried chipped	4,300	Olives—Greek	3,288
Bouillon—cubes or		Popcorn	1,940
powder	24,000	Pretzels	1,680
Caviar	2,200	Salad dressing—Italian	2,092
Cod—salt	8,100	Soy sauce	7,325
Herring—hard	6,231		

Any appreciable amounts of these edibles in the diet could raise the daily sodium intake to well over 20 grams. Looking at all of these figures makes it clear why it has been so incredibly difficult for people to remain on low-sodium diets. At the very least it involves a total renunciation of prepared foods, staying away from restaurants, and maintaining a spartan discipline.

Returning to the evolutionary viewpoint, it seems reasonable to assume that whenever a species radically alters its diet, metabolic upsets and their sequelae may be anticipated. This should be especially true with respect to essential nutrient requirement and mineral constituents that are unique. This argument has previously been used to favor a high vitamin C intake, and while it has certainly not been conclusively proven, it deserves serious consideration. We do not have any accurate method of measuring the time scale in genetic adaption to dietary changes, but we would certainly anticipate a very long period of selection that would involve lower vitality and lifespan for those unable to make the metabolic adjustment.

From a long term point of view, salt is probably a recent addition to the diet. In comparatively modern historical times, salt was introduced by the Europeans into India and parts of the Western Hemisphere. Persons living on high-milk, high-meat diets, such as nomadic herdsmen, do not eat salt with their food. The use of salt probably accompanied the shift from hunting to agricultural civilization. In the process, however, the total sodium intake has progressively risen to levels far above those maintained by non-salt-eating cultures. The remaining question is why has the sodium content of foods drifted to such high values in the contemporary American diet?

The two items under baby food in the second listing in this essay provide the clue. We grew up from infancy trained to expect a salty taste to foods. Modern infants will do better since some manufacturers, responding to this type of information, have reduced the amount of sodium chloride in baby foods.

Think of the number of people you know who salt their food before tasting it. As we become habituated to salt by constant exposure we gradually seem to add increasing amounts. In the end a diet of up to 20 grams a day of sodium is achieved by the use of other additives as preservatives, taste enhancers, and processing aids. A useful experiment is to go on a sodium-free diet for four full days and then return to normal prepared foods. For the first two or three meals ordinary foods will have a rather salty taste about them.

Well, interesting as the general evolutionary and cultural arguments may be, they do not constitute proof that the general levels of sodium in the diet are high enough to be toxic to the general population. The toxicity problem has recently been reviewed in the second edition (1973) of *Toxicants Occurring Naturally in Foods*, published by the National Academy of Sciences. The second chapter, written by George R. Meneely, sums up as follows: "There is no doubt that excess sodium is poisonous for many species of animal, that potassium exerts a protective effect, that there is a hereditary component in individual susceptibility to the hypertensigenic effect of salt, and that young animals are more sensitive than older ones. There is incontrovertible evidence that excess salt raises blood pressure in some humans, that salt restriction lowers blood pressure in some hypertensive patients, and that extra potassium is beneficial in some hypertensives. Beyond these solidly established facts, the human data are less clear-cut, but the weight of the evidence supports the transfer of the implications of the animal to man. Certainly if any newly proposed ingredient of diet or medicament evidenced such noxious effects in animal experiments, authorities would be compelled to view its promiscuous exhibition with grave alarm."

Given all of these factors, there exist strong enough presumptions to warrant a requirement for labelling of prepared foods for sodium content. That way the buyer can have the choice of purchasing low-sodium foods and more manufacturers would have the incentive to lower the addition of this alkali. The test for sodium involves a blender and a flame photometer. Excellent commercial instruments are available and are in general use in clinical laboratories. Sodium determination would not constitute a burden on the food processor.

On the other hand, the hypertensive could keep much better track of his sodium intake and could avoid high-sodium extremes even if not on a regulated diet. The rest of the population could have the option of deciding on the merit of evolutionary arguments. The simple device of noting the sodium content of prepared foods would be a big step toward truth in eating.

Molecular Cosmetology

I was in severe danger of acting like a male chauvinist. My hair was being trimmed by an attractive lady who inquired as to whether I wanted a treatment with a nucleic acid hair conditioner. My first impulse was to ask how they got the message in the medium. My second impulse was to humor her, but I decided that to do either would have been to demean her intelligence, so we entered into a serious discussion of the role and function of nucleic acids. The conversation and the haircut ended with a promise on my part to look into the possible role of nucleic acids in the care of hair and scalp. This led me to sections of the library I don't often frequent, not to mention my readings in journals such as The American Hairdresser Salon Owner. The results of these studies are summed up in the following letter:

Dear Lynn,

I must apologize for taking so long in sending you the information on nucleic acid in hair preparations, but I have been having such a hard time finding any solid facts (trade secrets, you know) that it has been necessary to work things out from first principles.

Starting about 1950, a theory has developed which is known as "the dogma of molecular biology." Briefly it states that genetic information is stored in DNA (deoxyribose nucleic acid). In growing cells this information is transferred to RNA (ribose nucleic acid), which acts as a messenger to specify the proteins that are produced. These molecules of protein are the active structural and functional components of living systems. Each species of plant and animal has a unique message encoded in its nucleic acids which determines its genetic properties.

Since nucleic acids function primarily in the making of protein, they would have to get deep within the skin to the hair-forming cells of the follicle in order to have any effect. DNA can act only at the gene level, and experiments involving gene transfer are so potentially dangerous that they must be carried out (by federal ruling) in special laboratories called P_2 and P_3 facilities. In any case, the various preparations probably contain yeast nucleic acid, and one would hardly want the scalp to be synthesizing yeast proteins. RNA, while not subject to government restrictions covering DNA recombination, would still have to get down into the scalp to the active cells in order to have any effect. It is very unlikely that such molecules penetrate that far, extremely improbable that they get into the cells, and dubious in the extreme that they would be functional if they managed to get in. Even assuming that all of these very low probability events occurred, I must repeat the question: Who would want yeast material growing out of his or her head? My suspicion is that the nucleic acids in hair preparations are so degraded that they could hardly be expected to have any biological effect at all.

What appears to be happening is that certain hair preparation manufacturers are capitalizing on the fact that the public has heard of nucleic acids but is unaware of the biochemical details of how they work. No one can expect a cosmetologist to be up on the latest developments in molecular biology. Therefore, you have to assume good faith on the part of your suppliers, and that good faith is not always justified. Thus when a manufacturer states, "Without RNA and DNA it would be impossible for the human system to form hair, skin, or nails," he tells the truth but deludes you into thinking that externally applied DNA or RNA will have something to do with hair, skin, or nails. To have the slightest chance of working this way, it would have to be human nucleic acid (where would they get it?), and then it would involve the most dangerous aspects of somatic genetic engineering. I suspect that what you are now getting is degraded yeast nucleic acid, which is both harmless and worthless, although I'll bet it's expensive.

While reading through advertisements on hair preparations in order to write this letter to you, I became intrigued by the fact that many shampoos and hair conditioners make claims that they contain protein. This seemed a bit unusual to me because in the laboratory we always refrigerate or freeze our protein preparations in order to stabilize them. These types of molecules have a great tendency to deteriorate on standing. My researches on this subject took me to the cosmetics section of the largest discount store in town, where I read labels for about 45 minutes while the saleslady kept eyeing me as if I were a shoplifter (we have to make great sacrifices in the cause of science). However, the results of the study made it all worthwhile. Now get a firm hold of your scissors before I tell you this one. None of those "protein" preparations contained any protein. In the list of ingredients on the back they all stated "hydrolyzed animal protein," which is simply not protein at all.

To make sense of all this we have to go back to some more basic biochemistry. Proteins are large molecules made up of building blocks that are amino acids. A protein possesses its specific properties because of the kinds and arrangements of the subunits. Hydrolyzed animal protein starts out as large molecules but is then treated with hot acid and broken down into its amino acid subunits. What the hair preparations contain is not protein but a collection of amino acids. To sell someone amino acids and maintain that it is protein is like selling someone a pile of bricks and maintaining that you are selling them a house.

Once again we see what the manufacturers have done.

Protein has become a popular word because of nutritionists (who also frequently misuse it—in nutrition it's the amino acids that count) and because of a general interest in science. Also, everyone is aware that hair is made of protein (then, again, so is botulinus toxin). So the idea is to put protein in hair preparations. But that is too difficult and expensive because the protein may denature and get sticky; so they first cut the large molecules up in little pieces, and it is no longer recognizable. There is not a biochemist around who would call the hydrolyzed material "protein." Amino acids go in the jar and "protein" goes on the label.

Now, what do amino acids do to the hair? I have been unable to find out. I could imagine that some sulfur-containing amino acids (cystine, cysteine) might have an effect, but that would involve quite complicated oxidation-reduction chemistry. Amino acids on the scalp could, of course, serve as food for skin bacteria, but I don't know how much of it stays around. I suspect that, by and large, it comes off in the rinse and down the drain. Thus the so-called protein in hair preparations is probably in the category of harmless, worthless, and expensive.

And while we're being in a cynical mood, I'd keep my eye out for vitamins, fatty acids, lecithins, minerals, and all other biochemicals in hair preparations. Until someone explains to me exactly what all these reagents do, I'm going to be from Missouri. Show me!

The problems we've been discussing are by no means unique to the hairdressing business. It's quite common to lift the language of science for commercial purposes without retaining fidelity to the spirit of scientific truth and accuracy. What can we do about it? I believe that an educated public with a deeper understanding of science is the only long-term solution. In the meantime, we'll have to keep calling them as we see them.

Thanks for the fine haircut.

Sincerely,
Harold